ECLIPSE

ECLIPSE

Bryan Brewer

For Pat Ehr—
Best wishes,
Bryan Brewer
Feb. 26, 1979

EARTH VIEW

Seattle, Washington

Library of Congress Cataloging in Publication Data

Brewer, Bryan.
　Eclipse.

　Bibliography: p.
　Includes index.
　1. Eclipses, Solar. I. Title.
QB541.B74　　523.7'8　　78-73047
ISBN 0-932898-27-0
ISBN 0-932898-26-2 pbk.

For information, contact:　Earth View, Inc.
　　　　　　　　　　　　　　1629 Madrona Drive
　　　　　　　　　　　　　　Seattle, WA 98122

Printed in the United States of America

Printed by Snohomish Publishing, Snohomish, WA 98290

———————————————

First Printing: November 1978
Second Printing: January 1979

To Molly

As in the soft and sweet eclipse,
When soul meets soul on lover's lips.

Shelley, Prometheus Unbound

Contents

Preface

I wrote this book for two reasons. First, I want to pass on the excitement I feel about seeing my first total solar eclipse on February 26, 1979. From all that I have read and heard from those who have witnessed totality, it is an experience to be cherished. For a brief moment we can step out of our ordinary world and into the realm of time and space on a truly cosmic scale. It is a chance to see a stunning view of the star at the center of our solar system and to gain some perspective on our place in the universe. And it's for everyone — no special equipment or skills are needed. As I say later in the book, "It's simply a matter of being in the right place at the right time — and knowing what to look for."

My other motive is to share information. I feel that the experience of a total solar eclipse ought to be balanced with an understanding of why it happens and how people have reacted to eclipses in the past, and I have organized this book with that purpose in mind. After some basic facts about eclipses are introduced, the first chapter traces humankind's fascination with this celestial phenomenon from Stone Age times to the present. Against that background chapter 2 explains what happens during an eclipse and how these events are repeated in time and space. The third chapter offers practical advice on getting out there and seeing an eclipse for yourself. (Although the February 26, 1979, eclipse is used for illustration, this chapter as well as the entire book is designed to have enduring value beyond this particular event.) Finally, the epilogue takes a peek at future solar eclipses, Halley's Comet, and extraterrestial communication.

Some notes of thanks are in order to a few of the many special people who have helped me in the creation of this book: First, to my wife, Molly, for just about everything, but most importantly for her trust in me and our future; to Dave Becker, for his unfailing encouragement and superb editing; to Dennis Schatz, the Pacific Science Center's Director of Astronomy Education, for technical advice and his friendly empathy with this project; to Joyce Andresen, for her helpful library research "with a smile"; to Eleanor Mathews, for her excellent maps; to Sydney Omarr, for his uncannily timed inspirations to persevere; to my son, Devin, who somehow managed to have his birthday on February 26; and loving gratitude for the special assistance from the Earth Gospel Players.

Bryan Brewer
Seattle, Washington

Foreword

The starry heavens are as much yours to enjoy as they are an astronomer's, which is a basic message of this informative book by Bryan Brewer. When we look outward with awe and wonder at the marvels awaiting us in an unlimited universe, it is well to recall what he tells us here about the roots of our ancient fascination with the sky. Even if you are someone merely making a wish upon the vision of a bit of cosmic rock burning itself out in Earth's atmosphere, you are in tune with that unbroken past.

Eclipses, those positive markers of our relative movement through the void, make a superb focal point for our outward vision. There is a direct line between Alan Shepherd driving a golf ball across the Moon's surface and the Celtic stonemasons whose stellar observatory still stands on the Salisbury plain. From Stonehenge to that first landing on the Moon, the human statement is plain to read: *If it moves, we want to know how and why.*

When you stand in the shadow of those bluestone menhirs at Stonehenge to mark the moment of the summer solstice, it is not difficult to feel an affinity with those early astronomers. The archaeological evidence that this carefully arranged series of stone circles was also used to predict eclipses appears to be conclusive. They accomplished much with little beyond their own muscles and intellectual inspiration.

That is likely to be the view distant future generations will take when they consider today's accomplishments. We are always primitives where our own far-off descendants are concerned.

Let the computer-written ephemeris tell you then where and when the next eclipse will occur; that prediction is not different in kind from what the Celtic astronomers told their people.

It is, however, different in quality. It is doubtful that you will perform some esoteric ritual to force the dragon (or snake or worm) to disgorge the Sun.

Mysteries remain, though, and we remake our mythology with increasing frequency, partly because we are still creatures of this planet and caught by a racial compulsion to penetrate beyond the regions that we have already mapped.

I like to think of Bryan Brewer's book as a map of the influences and rhythms contained in eclipses. It is well to remember that more than half the Earth's human population still uses astrology as a guide in the making of decisions. There is a possibility that a kernel of truth remains at the core of this ancient belief. We are Earth creatures. It would be remarkable if the rhythms that influence this planet where we evolved produced no effects on our flesh comparable to the influences upon our religions and philosophies. When we look at the heavens, we look at a cosmic clock that has marked every evolutionary development upon this mundane surface. That clock is still ticking, as the eclipse reminds us.

Frank Herbert
Port Townsend, Washington

Introduction

Into the Mouth of the Dragon

It's Monday morning, February 26, 1979 — the day of the eclipse. It has been scarcely half an hour since sunrise and the crisp chill of the winter air sends shivers through your body. As you move around to keep warm, you start wondering if you really want to be here. Crawling back into bed seems much more appealing than standing out here in the cold, miles from nowhere. Just as you're asking yourself "Is it all worth it?" you hear a shout of excitement.

"First contact! The eclipse has begun."

You rush over to where the eclipse viewer has been set up. People are looking at the image of the Sun projected onto the screen. And sure enough, there on one side of the bright disk, a tiny bit of the Sun has been covered up. You watch for a few minutes before you notice how it's changing. The Moon, slowly moving across the sky, is gradually blocking out more and more of the face of the Sun.

The timing of the eclipse was perfect, beginning within several seconds of the exact moment predicted for this location. Your group selected this site near the Washington-Oregon border months ago. You wanted a place near the center of the eclipse path for maximum duration of totality, with an unobstructed view of the morning Sun. After this one, there won't be another total solar eclipse visible from the United States for 38 years. And now, in a little less than an hour, all your plans and expectations will come to a climax in this once-in-a-lifetime event.

The feeling of excitement in the air begins to grow. Every few minutes you check the progress of the Moon inching across the image of the Sun. The Sun appears as a smaller and smaller crescent of light.

For most of the hour after the beginning of the eclipse you barely notice the decrease in sunlight. But as the time approaches for the beginning of totality, the landscape turns darker and darker. A growing sense of uneasiness seems to stir up twinges of an unspoken, primitive fear in all those present to witness this event.

1

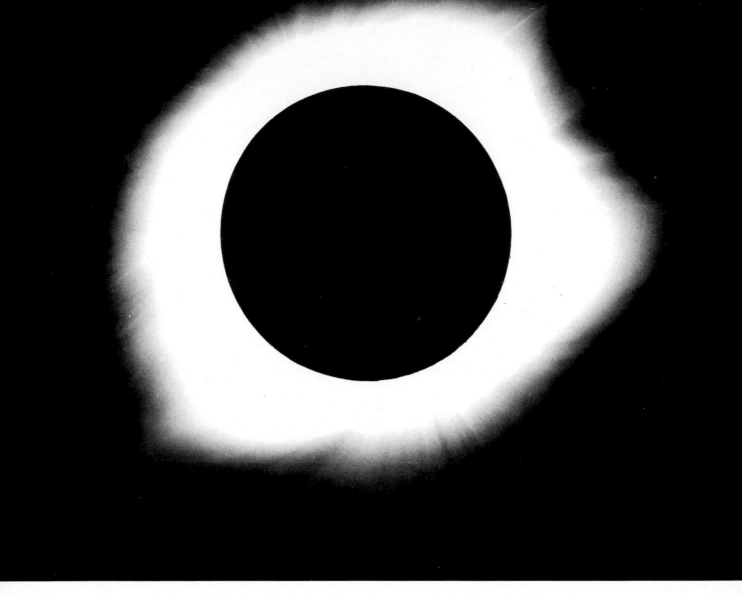

Eclipse of September 21, 1922, Wallal, Australia

As the narrow crescent of sunlight starts to disappear, little specks of light hang on for a few seconds more. And then, the dark shadow of the Moon rushes over you at incredible speed. The sky is suddenly dark and the corona surrounding the Sun bursts into view.

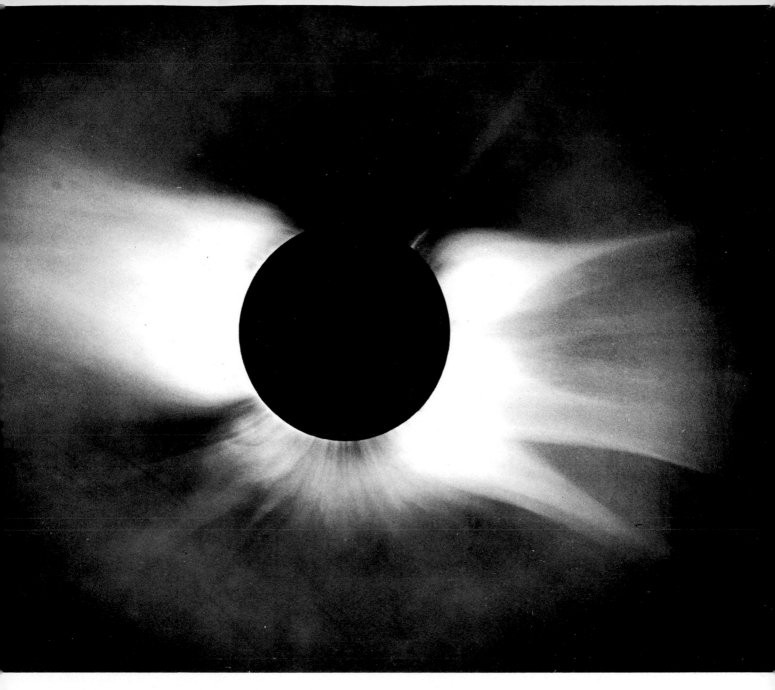

Eclipse of June 30, 1973, Kenya, Africa

Everything is silent. All eyes are held captive by this breathless spectacle in the sky. The irregular shape of the pearly white corona is spread behind the blackened Moon. Your eyes follow the wispy streamers of light extending out into the unreal darkness of the morning sky.

The sunlight from beyond the shadow casts a reddish glow near the horizon. You look around to notice how strangely things appear in the pale illumination of this eerie light. Birds have stopped singing; plants and animals react as if night has fallen. The sudden darkness seems to bring time and Nature to a quiet halt.

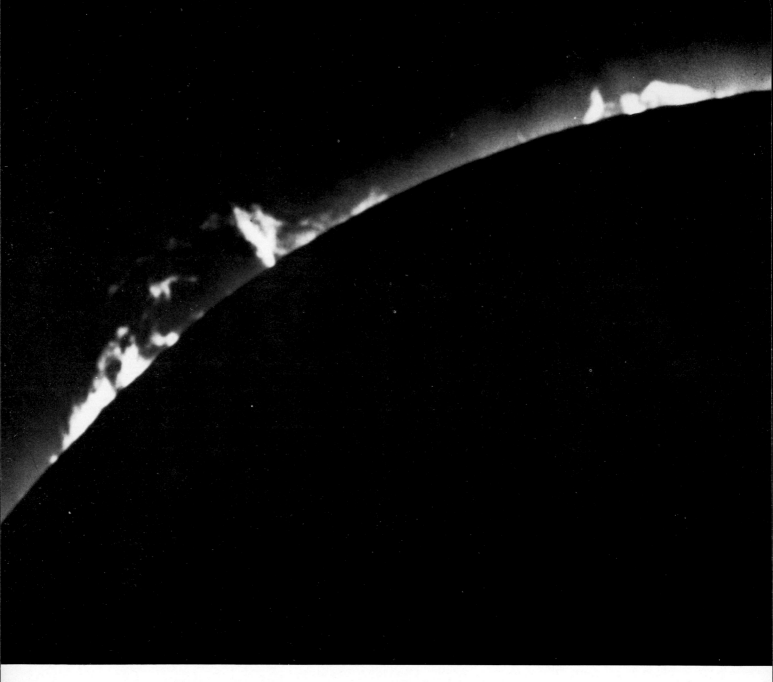

Solar prominences at the eclipse of August 31, 1932, Fryeburg, Maine

Precious seconds pass as you take in as much as you can of the beauty
of these special two minutes in time. It is dark enough to see some stars
and possibly a planet or two. Perhaps a few bright red solar prominences rise from the surface of the Sun, these arching flamelike erup-
tions punctuating the aura of light shining around the dark disk of
the Moon. But one thing dominates the sky. The delicate corona — the
halo of our Sun — shows its glory for these brief moments, giving you a
vision never to be forgotten.

And then, as suddenly as it began, it's over. The shadow passes on and the sunlight returns. The excitement of totality is replaced by a soothing calm. No one talks much at first. As the Moon gradually uncovers more and more of the morning Sun, you are quiet. You want to savor the freshness of the experience, at a loss for words to explain it. Yet somewhere deep inside you have a feeling. An unexpected intimacy with the awesome and relentless forces of Nature has somehow become yours. And you sense that you may never feel quite the same again.

The darkening of the Sun in the middle of the day will always seem an unnatural event. Even today with our scientific understanding of the Earth and space, it still can't help but seem a little frightening to watch the Sun disappear and leave us shrouded in midday darkness. Fortunately, the Sun's "abandoning" of the sky during an eclipse is only temporary; daylight returns to reassure us just as it has done for eclipses throughout the ages. (The word "eclipse" comes from the Greek word meaning "abandonment.") But the Greeks were not the first civilization to leave us with a history of observing eclipses.

The ancient Chinese document *Shu Ching* contains the earliest record of a solar eclipse. Most historians agree that the date was October 22, 2134 B.C., when "the Sun and Moon did not meet harmoniously." The story goes that the two royal astronomers, Hsi and Ho, had neglected their duties and failed to predict the event. Widespread Oriental belief held that an eclipse was caused by an invisible dragon devouring the Sun. Great noise and commotion (drummers drumming, archers shooting arrows into the sky, and the like) were customarily produced to frighten away the dragon and restore daylight. When this eclipse took place, the emperor was caught unprepared. Even though the Sun returned, the angry ruler ordered the astronomers beheaded!

Peruvian natives reacting to total solar eclipse in ancient times

The people of many cultures have believed that an eclipse is an omen of some natural disaster or the downfall of a ruler. During eclipses in India people immerse themselves in water up to their necks, believing this act of worship will help the Sun and Moon defend themselves against the dragon. In Japan, the custom is to cover wells during an eclipse to prevent poison from dropping into them from the darkened sky. And as recently as the last century, the Chinese Imperial Navy fired its ceremonial guns during an eclipse to scare off the invisible dragon.

This ominous view of eclipses is not the only one. In Tahiti, for example, eclipses have been interpreted as the lovemaking of the Sun and the Moon. Even to this day, the Eskimos, Aleuts, and Tlingits of Arctic America believe an eclipse shows a divine providence: the Sun and the Moon temporarily leave their places in the sky and check to see that things are going all right on Earth. But regardless of the meaning given to them, eclipses will continue to occur, always obeying the regular timetables of celestial motions.

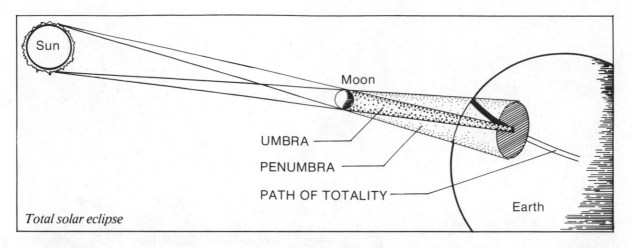

UMBRA

PENUMBRA

PATH OF TOTALITY

Total solar eclipse

The Moon travels in its orbit around the Earth once a month. An eclipse of the Sun takes place whenever the new Moon, passing right between the Sun and the Earth, casts its shadow on our planet's surface. (This doesn't happen every month; usually the Moon passes slightly above or below the line between the Sun and the Earth.) From anywhere within the dark cone of the complete shadow, called the *umbra* (from Latin for "shade"), the Sun will appear in total eclipse. When viewed from the area covered by the larger cone of the *penumbra* (literally, "almost shade"), the Sun is seen in partial eclipse.

The partial phases of a solar eclipse are normally visible from a broad area of the Earth as wide as 5,000 miles. A total eclipse, on the other hand, can be seen only from a narrow band called "the path of totality." This path, sometimes as wide as 200 miles, covers only about one-half of one percent of the Earth's surface. And because there are fewer than 70 total eclipses per century, the chance to see one is for most of us a once-in-a-lifetime event.

It is remarkable that total solar eclipses even occur at all. They are possible because the Sun and the Moon appear from Earth to be about the same size in the sky. The Sun, whose diameter is 400 times that of the Moon, happens to be about 400 times as far away from the Earth. This condition permits the Moon to just barely cover up the Sun. In fact, if the Moon's diameter (2,160 miles) were just 140 miles less, it would not be large enough to ever completely cover the Sun: a total solar eclipse could never happen anywhere on Earth!

Fortunately they do occur. Once every year and a half (on the average) Nature permits someplace on the Earth a view of the beautiful corona surrounding the Sun. For ancient people this spectacle of the sky must have had a forceful effect on their consciousness. From the beginnings of written history humankind has recorded eclipses and attempted to understand them. And now, recent investigations have pushed that history further back in time: the intelligent plan of an ancient stone monument, whose construction began as early as 2400 B.C., shows that it could have been used for eclipse predictions.

Chapter 1

Eclipses Throughout the Ages

Stonehenge: Eclipse Computer?

Every year on the first day of summer, the Sun rises at a point that is
farther north than on any other day of the year. At the ruins of Stonehenge in England, this solstice sunrise appears on the horizon in direct
alignment with the massive heel stone. This is the most outstanding
feature of this ancient monument, built during the same era as the Great
Pyramid of Egypt. There is little doubt that the builders of Stonehenge
used it to mark this special day as the beginning of each year. By
counting the number of days between these annual alignments they
could determine the length of the year. This could serve as a practical
calendar to mark holidays and seasonal festivals and to ensure the
timely planting and harvesting of crops.

But to predict eclipses, knowledge of two other cycles is required. One
of these — the length of the lunar month — is easily determined. It is
simply the number of days between one full Moon and the next. This
cycle of 29½ days is marked at Stonehenge by two rings of 29 and 30
holes, which together average 29½. The other cycle, however, is of an
altogether different character: it is a cycle of rotation of two invisible
points in space. The evidence shows that the builders of Stonehenge
probably discovered this cycle and could have used it to predict eclipses.

These two invisible points in space are called the lunar *nodes* (from the
Latin for "knot"). They are the points where the Moon's orbit, which
is tilted at a slight angle, intersects the plane of the Earth's orbit. It
would have taken many decades of watching countless risings and
settings of the Moon to figure out the cycle of the lunar nodes. This
information — which must have been passed on from generation to
generation — is preserved at Stonehenge. All the Moon alignments
necessary for determining this cycle are marked by massive stones.

Opposite: View to the northeast where the Sun rises on the first day of summer at Stonehenge

Ruins of Stonehenge today

Who were these people who observed this subtle cycle even before the first metal tools were used by humankind? Some have suggested that Stonehenge was built by Druids, but we don't really know much about the builders. We do know that the actual motions of the Sun and the Moon are reflected in the structure of Stonehenge, and we can reason how it may have been used to keep track of these cycles. The number of stones or holes in the ground in the various rings around Stonehenge each represents a certain number of days or years in the cycles. By moving markers (such as stones) around a ring in time with the cycles, the positions of the Sun and Moon — and the two invisible points — can be tracked. (The details of this method are explained in chapter 2.)

An eclipse can occur only when the Sun is close to being aligned with a node. By using Stonehenge to keep track of the position of the Sun and the nodes, these "danger periods" for eclipses can be predicted. A new (or full) Moon appearing during one of these periods would call for a special vigil to see if the solar (or lunar) eclipse would be visible from Stonehenge. (A lunar eclipse occurs when the full Moon is engulfed in the Earth's shadow.) A total solar eclipse would be a rarity. But the law of averages reveals that either a partial solar eclipse or a lunar eclipse can be seen (weather permitting) from the same point on the Earth about once every year.

Why would eclipses have been so important to these ancient people of Stonehenge? Perhaps they considered the darkening of the Sun or the Moon a fearsome event — a celestial omen of doom or disaster. Many cultures have interpreted eclipses this way. The sophistication of the astronomy of Stonehenge suggests that the builders had something different in mind. Their understanding of the solar and lunar cycles must have led to a high regard for the cosmic order. Eclipses may have been seen as affirmations of the regularity of these cycles. Or perhaps the unseen lunar nodes formed an element of their religion as invisible gods capable of eclipsing the brightest objects in the heavens.

The idea that Stonehenge may have been a center for some kind of worship has occurred to many of its students. It is not hard to imagine Stone Age people gathering at a "sacred place" at "sacred times" (such as solstices, equinoxes, and eclipses) to reaffirm their religious beliefs through ritual practices. British antiquarian Dr. William Stukeley, who in 1740 was the first to note the summer solstice alignment at Stonehenge, advanced the notion that the monument was built by Druids to worship the serpent. He claimed that Stonehenge and similar stone circles had been serpent temples, which he called "Dracontia." Could this serpent symbolism be related to eclipses? Recall that the key to eclipses is the position of the lunar nodes. The length of time for the Moon to return to a node (about 27.2 days) astronomers call the *draconic month.* (Draco is the Latin word for "serpent" or "dragon.") Perhaps the mythical serpents of Stonehenge and the legendary dragon that eats the Sun are symbols of the same thing: the invisible presence in time and space that eclipses the Sun and the Moon.

Fanciful illustration of early Stonehenge celebrations

11

The Birth of Astronomy

Whatever the reasons for Stonehenge, they are lost in time; the builders left no written records. During this same period in history (between 3000 B.C. and 2000 B.C.), study of the heavens was developing as a written science in the Middle East. Astronomers in Babylonia and Assyria kept track of time by carefully observing the motions of the Sun and the Moon. They increased the accuracy of their measurements by recording the details of solar and lunar eclipses. As they studied this record of centuries of eclipses, a pattern of repetition began to emerge: eclipses tend to repeat themselves every 18 years, although they recur at different places on the globe. This eclipse cycle, called the *saros*, is used even to this day to make predictions. For example, the February 26, 1979, eclipse is included in a *saros* series. On February 15, 1961, exactly 18 years and 11 days earlier, a solar eclipse took place. Another solar eclipse will occur on March 9, 1997, again after 18 years and 11 days.

The Babylonian discovery of the *saros*, important for eclipse predictions, is not the most famous of their contributions to astronomy. As early as 3000 B.C., they originated the division of the sky into the 12 signs of the zodiac, and the names they gave to each sign are still used. Today these names and symbols (the symbols are of unknown origin) are more familiar to the practice of astrology; but in ancient Babylonia, astronomy and astrology were inseparably connected.

The religion of the Babylonians was based on the belief that earthly affairs were influenced by the motions of heavenly bodies. It was the duty of the astrologer-priests to keep watch on the skies and warn of any disasters that might be signaled. They developed an elaborate system of celestial omens to "divine" the future. It seems that the Babylonians were more interested in observing the paths of the Sun, the Moon, and the planets across the background of the stars than in discovering the true nature of the world around them. They also believed that each sign of the zodiac was influenced by one of the "ruling planets," which imparted its qualities to events related to that sign.

The Babylonians also contributed to the establishment of the seven-day week. As the 12 signs of the zodiac related to the 12 lunar months in a year, the seven days in a week probably come from the quarter phases of the lunar month. Also, ancient astronomers/astrologers recognized seven planets (including the Sun and the Moon); they associated each planet, personified as a celestial god, with a day of the week. The present names for the days of the week are derived from this same scheme, using Roman or Norse names for the planetary gods.

ZODIAC SIGN	SYMBOL	ANCIENT "RULING PLANET"
Aries	♈	Mars
Taurus	♉	Venus
Gemini	♊	Mercury
Cancer	♋	Moon
Leo	♌	Sun
Virgo	♍	Mercury
Libra	♎	Venus
Scorpio	♏	Mars
Sagittarius	♐	Jupiter
Capricorn	♑	Saturn
Aquarius	♒	Saturn
Pisces	♓	Jupiter

Planetary gods from which the names of the days of the week originated:

1. *Sunday* . . . . *(Sun)*
2. *Monday* . . . . *(Moon)*
3. *Tuesday* . . . . Tiw, *Norse God of War (Mars)*
4. *Wednesday* . . . . Woden, *Norse Chief God (Mercury)*
5. *Thursday* . . . . Thor, *Norse God of Thunder (Jupiter)*
6. *Friday* . . . . Freya, *Norse Goddess of Marriage (Venus)*
7. *Saturday* . . . . *(Saturn)*

The Winged Sun Over Egypt

As the Babylonians were developing the science of astronomy, the ancient Egyptian civilization was flourishing. Pyramids, temples, and tombs attest to the high state of development of their art and technology. They measured the length of the year by observing the rising of Sirius, the brightest star in the sky. The Great Pyramid at Giza is aligned to the four points of the compass; it was built with a passage-way in alignment with the star that was then the pole star, Alpha Draconis. There is no doubt that the Egyptians watched the heavens. The clear skies of the Nile Valley were ideally suited for celestial observation. Yet no one has found a single reference to an eclipse, either of the Sun or the Moon, in all of ancient Egyptian history.

This apparent gap in Egyptian astronomy has puzzled many historians. Was Egypt shortchanged on total solar eclipses? Far from it. The accompanying map shows all the paths of totality across the Nile Valley in the second and third millenia B.C. The solar corona was visible from somewhere in ancient Egypt during this period on an average of once every 75 years. It is hard to imagine that the spectacular recurrence of total solar eclipses could go unrecorded, especially by a culture that so worshipped the Sun.

TOTAL SOLAR ECLIPSES IN THE NILE VALLEY (3000 B.C. - 1000 B.C.)

Mar. 23, 2861 B.C.	Apr. 20, 2044 B.C.
Nov. 19, 2837 B.C.	Sep. 15, 1884 B.C.
Apr. 1, 2471 B.C.	Dec. 21, 1741 B.C.
Sep. 2, 2469 B.C.	Apr. 16, 1699 B.C.
July 25, 2430 B.C.	May 9, 1533 B.C.
Oct. 27, 2379 B.C.	June 1, 1478 B.C.
June 25, 2354 B.C.	May 14, 1338 B.C.
Mar. 23, 2340 B.C.	July 27, 1258 B.C.
Dec. 20, 2289 B.C.	Aug. 19, 1157 B.C.
June 29, 2159 B.C.	Feb. 14, 1129 B.C.
Sep. 11, 2079 B.C.	July 31, 1063 B.C.

(Map shows central lines of total eclipses. All dates are B.C.)

14

Perhaps the view of totality was preserved in symbolic form. The solar corona has a distinctive appearance during some eclipses. The size and shape of this halo around the Sun varies over a cycle of 11 years. (This is the sunspot cycle explained later in this chapter.) During the minimum phase of this cycle, the brightness of the corona is less intense, but extending to either side are long streamers of light. Because these equatorial streamers are so faint, they are difficult to photograph. Yet in clear skies they are plainly visible to the naked eye.

The similarity of these eclipse streamers to the symbolic wings of the Egyptian Sun is clear. The drawing above was made by Samuel P. Langley from the summit of Pike's Peak during the eclipse of July 29, 1878. The long equatorial streamers are well defined. The symbol below is the winged disk of the Sun; it was one of the earliest solar representations in Egypt. It appears above the entrances of many tombs and temples and is said to commemorate the victory of light over darkness. Sometimes the symbol includes the heads of two serpents and the horns of a goat, also solar symbols.

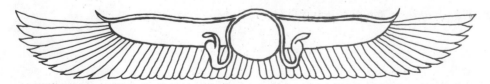

Could this view of the eclipsed Sun be the ancient source of this widespread symbol? English astronomer E.W. Maunder put it this way: *

> ...there can be little doubt that the Sun was regarded partly as a symbol, partly as a manifestation of the unseen, unapproachable Divinity. Its light and heat, its power of calling into active exercise the mysterious forces of germination and ripening, and the universality of its influence, all seemed the fit expressions of the yet greater powers which belonged to the Invisible.

> What happened in a total solar eclipse? For a short time that which seemed so perfect a divine symbol was completely hidden. The light and heat, the two great forms of solar energy, were withdrawn, but something took their place. A mysterious light of mysterious form, unlike any other light, unlike any other single form, was seen in its place. Could they fail to see in this a closer, a more intimate revelation, a more exalted symbolism of the Divine Nature and Presence?

Winged solar disk adorns the top of King Tutankhamun's ceremonial chair.

Knowledge, vol. XX, p.9, January 1897.

Eclipses in History and Literature

Stonehenge, Babylonia, Egypt — each culture developed a unique approach to eclipses. But only the Babylonians discovered the long-range prediction cycle, the *saros*. An eclipse cycle can also be used to go backward in time. This technique has proven useful to historians in fixing exact dates of past events.

Numerous systems were used in ancient civilizations to keep track of the passage of time. Typically, routine happenings would be recorded as so many days, months, or years after some memorable event such as the crowning of a ruler, a natural catastrophe, or other momentous occasion. Often there would be no indication of exactly when the reference event took place. If an eclipse was described in the record of events, it could be compared with actual eclipses that were known to have happened near the time and place in question. If there were only one eclipse that fit the description, then the dates could be fixed with certainty. Many historical chronologies have been verified or compared using this method.

The earliest record of a solar eclipse comes from ancient China. The date of this eclipse, usually given as October 22, 2134 B.C., is not certain. Historians know the account was written sometime within a period of about two hundred years. During that time there were several total eclipses visible in China. The 2134 B.C. eclipse is simply the best guess.

The date of an eclipse referred to in the Bible is known for certain: " 'And on that day,' says the Lord God, 'I will make the Sun go down at noon, and darken the Earth in broad daylight.' " (*Amos* 8:9) "That day" was June 15, 763 B.C. The date of this eclipse is confirmed by an Assyrian historical record known as the *Eponym Canon*. In Assyria, each year was named after a different ruling official and the year's events were recorded under that name in the *Canon*. Under the year corresponding to 763 B.C., a scribe at Nineveh recorded this eclipse and emphasized the importance of the event by drawing a line across the tablet. These ancient records have allowed historians to use eclipse data to improve the chronology of early Biblical times.

Battle between Lydians and Medes (585 B.C.) halted by total solar eclipse.

What is probably the most famous eclipse of ancient times ended a five-year war between the Lydians and the Medes. These two Middle Eastern armies were locked in battle when "the day was turned into night." The sight of this total solar eclipse (the date is fixed as May 28, 585 B.C.) was startling enough to cause both nations to stop fighting at once. They agreed to a peace treaty and cemented the bond with a double marriage. The eclipse was predicted by Thales, the celebrated Greek astronomer and philosopher, but the prediction was probably not known to the warring nations.

The lunar eclipse of August 27, 413 B.C., had a different effect on the outcome of battle in the Peloponnesian War. The Athenians were ready to move their forces from Syracuse when the Moon was eclipsed. The soldiers and sailors were frightened by this celestial omen and were reluctant to leave. Their commander, Nicias, consulted the soothsayers and postponed the departure for twenty-seven days. This delay gave an advantage to their enemies, the Syracusans, who then defeated the entire Athenian fleet and army, and killed Nicias.

The spectacle of an eclipse, which had a powerful effect on decisions in battle, was equally impressive to ancient poets. A fragment of a lost poem by Archilochus contains the words:

> Nothing there is beyond hope, nothing that can be sworn impossible, nothing wonderful, since Zeus, father of the Olympians, made night from mid-day, hiding the light of the shining Sun, and sore fear came upon men.

This has been identified as a description of the total solar eclipse of April 6, 648 B.C. Another eclipse reference (from the Bible) goes like this:

> And I behold when he had opened the sixth seal, and lo, there was a great earthquake; and the Sun became black as sackcloth of hair, and the Moon became as blood.
>
> Revelation 6:12

This compelling passage is only one of a number of literary and historical connections between eclipses and earthquakes. The Greek historian Thucydides, in writing about the Peloponnesian War, remarked about "earthquakes and eclipses of the Sun which came to pass more frequently than had been remembered in former times." On another occasion he noted "...there was an eclipse of the Sun at the time of a new Moon, and in the early part of the same month an earthquake." Another Greek writer, Phlegon, reported the following events:

> In the fourth year of the 202nd Olympiad, there was an eclipse of the Sun which was greater than any known before and in the sixth hour of the day it became night; so that stars appeared in the heaven; and a great earthquake that broke out in Bithynia destroyed the greatest part of Nicaea.

This interest in linking these two types of events by coincidence may have been attempts to derive some order out of the unpredictability of earthquakes, possibly a carryover from the celestial omens of the Babylonians. Oddly enough, this type of coincidence seems to persist. The earthquake in Iran on September 16, 1978, the most devastating one of the year up to that time and which killed more than 25,000 people, occurred just 3½ hours before a total lunar eclipse was visible there!

These kinds of ominous events have played important parts in human history. When English poet John Milton, in *Paradise Lost*, wrote these lines

> As when the Sun, new risen,
> Looks through the horizontal misty air,
> Shorn of his beams, or from behind the Moon,
> In dim eclipse, disastrous twilight sheds
> On half the nations and with fear of change
> Perplexes monarchs

he may have been thinking of Charlemagne's son, Emperor Louis. This European ruler was so "perplexed" by the five minutes of totality he witnessed during the eclipse of May 5, 840, that he died (some say of fright) shortly thereafter. The fighting for his throne ended three years later with the historic Treaty of Verdun, which divided Europe into the three major areas we know today as France, Germany, and Italy.

Solar eclipses perplexed the common people as well. Medieval historian Roger of Wendover reported on the total eclipse of May 14, 1230, which occurred early in the morning in Western Europe: "...and it became so dark that the labourers, who had commenced their morning's work, were obliged to leave it, and returned again to their beds to sleep; but in about an hour's time, to the astonishment of many, the Sun regained its usual brightness." This was during the Dark Ages and an understanding of eclipses was not common knowledge.

Mark Twain used this ignorance of eclipses as an element of the plot in *A Connecticut Yankee in King Arthur's Court.* The hero of the novel, Hank Morgan, is mysteriously transported backward in time to Medieval England. He finds himself ready to be burned at the stake on a day when he knows a solar eclipse will occur. He "foretells" the event, claiming to have magical powers over the Sun. "The rim of black spread slowly into the Sun's disk,...the multitude groaned with horror to feel the cold uncanny night breezes... and see the stars come out..." Morgan promises to restore the daylight in exchange for his freedom. King Arthur agrees and, of course, the Sun returns. Twain gives the date of the eclipse as June 21, 528; this, however, is literary fiction. No such eclipse took place on or near that date.

A similar sort of deception was actually used by Christopher Columbus during his fourth voyage to the Americas. In 1503, he found himself stranded on the island of Jamaica, his ships damaged beyond repair and his provisions running low. At first he and his crew were able to get food from the natives in trade for baubles and trinkets. But as months passed without rescue, the Jamaicans finally refused to supply any more food. Faced with the prospect of starvation, the great Spanish admiral conceived an ingenious plan.

Columbus knew from his navigational tables that a total eclipse of the Moon would occur on February 29, 1504. He arranged a meeting with the natives that evening to coincide with the beginning of the eclipse. He announced that because God didn't like the way the natives were treating him and his crew, the Almighty had decided to permanently remove the Moon as a sign of his displeasure! Columbus timed his theatrics precisely; no sooner had he proclaimed the Moon's disappearance than the Earth's shadow began to steal across the face of the full Moon.

The natives were terrified. As the light of the Moon faded they pleaded with Columbus to restore it; they would give him all the food he wanted if he would bring back the Moon. Columbus told them he would have to retire to confer with God, which in this case was an hourglass timing the eclipse. Just before the end of the total phase he announced that God had pardoned them and would allow the Moon to return to its place in the sky. And as Columbus knew it would anyway, the Moon reappeared. The grateful natives resumed the supply of food, and Columbus and his crew were eventually rescued and returned to Europe.

Terrified Jamaicans plead with Columbus to restore eclipsed Moon.

The Science of Prediction

Today's astronomers are able to predict the precise time and location of solar eclipses. With advance notice of the event and a higher level of scientific understanding among people, there is no need for anyone to be frightened by what should be a marvelous experience of the beauty of Nature. In fact, many people make plans to travel to locations in the path of the Moon's shadow just to witness the spectacle of a total solar eclipse. Astronomers at the U.S. Naval Observatory in Washington, D.C., process the data and publish predictions a year or two in advance for all eclipses. The times and locations (including maps) appear in the annual publication *The American Ephemeris and Nautical Almanac.* According to the Observatory astronomer who maintains the computer programs for eclipses, the prediction of the path of totality is accurate to within one or two miles and the timing of the eclipse to within a few seconds.

Eclipse predictions have not always been that accurate. Stonehenge could be used to forecast the day of an eclipse, but not the specific time or place. The ancient Chinese, Babylonians, and Greeks made improvements, but it was not until the 17th century A.D., after Copernicus had shown that the Sun is at the center of the solar system and Newton had formulated the laws of gravity, that eclipse predictions achieved modern accuracy. The actual motion of the Moon is fraught with numerous small inequalities and discrepancies. In 1693 British astronomer Edmond Halley (of comet fame) was the first to notice a small but steady change in the Moon's motion called *secular acceleration.* This simply means that the Moon is slowly gaining speed in its orbit. Modern astronomers use ancient eclipse records, some several thousand years old, to determine the value of this change.

In 1824 a great practical stride was made in eclipse predictions: Prussian astronomer Friedrich Bessel introduced a group of mathematical formulas that greatly simplified the calculation of the positions of the Sun, Moon, and Earth. These "Bessel functions," which are used even today, laid the foundation for a monumental book on eclipses. An Austrian astronomer named Theodor von Oppolzer organized the calculation of all eclipses from 1207 B.C. to 2161 A.D. One year after his death in 1886, his *Canon of Eclipses* was published. It contains the details of the time and place for the 13,200 solar and lunar eclipses for those 34 centuries. Another Austrian astronomer, Friedrich Ginzel, used these data for historical research on eclipses. In 1899 he published his *Special Canon of Solar and Lunar Eclipses,* which shows the references in classical literature to all eclipses between 900 B.C. and 600 A.D. Both of these works have been valuable tools for historians who use eclipses to verify dates in history.

Opposite: Map produced by Edmond Halley showing path of totality across England on April 22, 1715

A Description of the Passage of the Shadow of the Moon, over England, In the Total Eclipse of the SUN, on the 22.Day of April 1715 in the Morning.

THE GERMAN SEA

SCOTLAND

THE IRISH SEA

IRELAND

St. GEORGE'S CHANNEL

THE CHANNEL

FRANCE

I. Wight

Engrav'd by John Senex

The like Eclipse having not for many Ages been seen in the Southern Parts of Great Britain, I thought it not improper to give the Publick an Account thereof, that the suddain darkness, wherein the Stars will be visible about the Sun, may give no Surprize to the People, who would, if unadvertized, be apt to look upon it as Ominous, and to Interpret it as portending evill to our Sovereign Lord King George and his Government, which God preserve. Hereby they will see that there is nothing in it more than Natural, and no more than the necessary result of the Motions of the Sun and Moon; And how well those are understood will appear by this Eclipse.

According to what has been formerly Observed, compared with our best Tables, we conclude the Center of the Moons shade will be very near the Lizard point, when it is about 5 min: past Nine at London; and that from thence in Eleven minutes of Time, it will traverse the whole Kingdom, passing by Plymouth, Bristol, Glocester, Daventry, Peterborough & Boston, near which it will leave the Island: On each Side of the Tract for about 75 Miles, the Sun will be Totally darkned; but for less & less Time, as you are nearer those limits, which are represented in the Scheme, passing on the one side near Chester, Leeds, and York; and on the other by Chichester, Gravesend, and Harwich.

At London we compute the Middle to fall at 13 min: past 9 in the Morning, when its dubious whether it will be a Total Eclipse, or no; London being so near the Southern limit. The first beginning will be there at 7 min: past Eight, and the end at 24 min: past nine. The Ovall figure shews the space the Shadow will take up at the time of the Middle at London; And its Center will pass on to the Eastwards, with a Velocity of nearly 30 Geographical Miles in a min: of Time.

N.B. The Curious are desired to Observe it, and especially the duration of Total Darkness, with all the care they can; for thereby the Situation and dimensions of the Shadow will be nicely determind; and by means thereof we may be enabled to Predict the like Appearances for the future, to a greater degree of certainty than can be pretended to at present, for want of such Observations.

By their humble Servant Edmund Halley.

Fig. W.

Modern Eclipse Expeditions

Astronomers today travel to all parts of the globe to gather eclipse data, chasing the Moon's shadow wherever it happens to touch the Earth. This has not always been the case. It is only in the last 150 years or so that eclipse expeditions have been in vogue. A notable exception occurred for the total solar eclipse of October 27, 1780: Samuel Williams, professor at Harvard, led an eclipse expedition to Penobscot Bay, Maine. The exceptional part of the story is that this happened during the Revolutionary War and Penobscot Bay lay behind enemy lines. Fortunately, the British granted the party safe passage, citing the interest of science above political differences.

Until the middle of the last century, most of the scientific interest in eclipses concerned the precision of orbital motion. Astronomers used the data from eclipse observations to refine their knowledge of celestial mechanics, which in turn led to more accurate eclipse predictions. Little attention was paid to describing the visible phenomena of total solar eclipses.

This situation changed when the Moon's umbra crossed populated parts of Southern Europe on July 8, 1842. Those who observed totality on that day were rewarded with a magnificent view of the corona and prominences. Francis Baily, English amateur astronomer, was the first to use the word "corona" as an astronomical term in describing this eclipse. Scientists were stirred to discover more about this halo of light and the "red flames" that appeared around the Moon. Luckily, the technology of photography was beginning to develop at that time. The first successful photograph of the corona was taken at the total eclipse in Northern Europe in 1851. Scientists in 1860 used eclipse photos taken in Spain to show that the solar prominences were definitely part of the Sun and not of the Moon as some had believed.

Opposite: 17th century astronomer Hevelius observing a solar eclipse by projection into a darkened room (from his Machina Coelestis, *1673)*

Corona seen in 1842

About this time another breakthrough happened that laid the foundation for the future of solar physics. For many years scientists had noticed a number of thin dark lines in the rainbow spectrum of light from the Sun. In 1859, German physicist Gustav Kirchhoff accounted for their origin: the lines occurred because of the chemical elements present in the Sun. Since each element has its own distinctive set of lines, the chemical composition of the Sun could be derived from its spectrum.

This new technique, called spectroscopy, was first applied to the eclipsed Sun on August 18, 1868. By this time, eclipse expeditions to remote areas of the globe were routine, and many traveled to India and Malaya to see this eclipse. British astronomer Norman Lockyer trained his spectroscope on the solar prominences and discovered a spectral line of a new chemical element. He named it helium (from Greek *helios*, the Sun); this familiar gas was not identified on Earth until 1895. At the same eclipse, he and French astronomer Pierre Jules Janssen, each working independently, figured out a spectroscopic method for observing the prominences without an eclipse.

19th century spectroscope used for studying Sun during eclipses

In the following ten years steady advances were made in the spectroscopy and photography of eclipses. Expeditions to America, the Mediterranean, India, South Africa, and Siam yielded new information about the composition of the Sun and the structure of the corona. These expeditions to remote areas presented many challenges to astronomers. Transporting large, sensitive telescopes and other instruments compounded the hardships of global travel a century ago. And once they were set up at a site within the path of totality, cloudy skies during the eclipse could defeat the purpose of the journey.

The expedition to India for the eclipse of December 12, 1871. These British astronomers (Norman Lockyer is seated at the left under the umbrella) made their observations from atop a tower at the old fort at Bekul. A crowd of astonished natives gathered around the tower; in their alarm at the sight of the disappearing Sun, the terrified people kindled a fire in preparation for a sacrifice. The astronomers, fearful the smoke would obscure their view, had the police stop the attempted fire-lighting.

High on her speculative tower
Stood Science waiting for the hour
When Sol was destined to endure
That darkening of his radiant face
Which Superstition strove to chase,
Erewhile, with rites impure.

Wordsworth
The Eclipse of the Sun, 1820

HARPER'S WEEKLY

A
JOURNAL OF CIVILIZATION

Vol. XXII.—No. 1130.]   NEW YORK, SATURDAY, AUGUST 24, 1878.   [WITH A SUPPLEMENT. PRICE TEN CENTS.

Entered according to Act of Congress, in the Year 1878, by Harper & Brothers, in the Office of the Librarian of Congress, at Washington.

28

None of these inconveniences deterred the scientists who traveled to Colorado and Wyoming for the total eclipse of July 29, 1878. The transcontinental railroad had been completed nine years earlier and astronomers were offered half-price fares for the trip from the East Coast. For a short period that summer, obscure towns in the West became centers of scientific activity. Famous astronomers from Europe and all over America turned out to see the eclipse in the clear skies of the Rocky Mountains. Even Thomas Edison (no astronomer himself) was there to test a new invention he claimed could measure the heat of the corona.

A group headed by Samuel P. Langley, later director of the Smithsonian Institution, climbed to the summit of Pike's Peak in Colorado to witness totality on July 29, 1878. The day before the eclipse was ominous; they experienced hail, rain, sleet, snow, and fog! But eclipse day was clear and their perseverance was rewarded by a startling sight: two coronal streamers extending in opposite directions as far as twelve diameters of the Sun. This was much wider than had ever been seen before by scientists.

Although the 1878 corona was very wide, it was actually not as bright as those seen in 1870 and 1871. Astronomers began to suspect that the corona's shape and intensity were related to levels of activity on the Sun. One of the measures of solar activity is the occurrence of sunspots. These dark blotches appear on the Sun's surface, sometimes lasting for many weeks. Some years earlier Heinrich Schwabe, a German amateur astronomer, noted that the average number of sunspots per day varied in a regular cycle of approximately 11 years. Could this cycle be linked to the changing shape of the corona?

Opposite: The "Great Solar Eclipse" of July 29, 1878

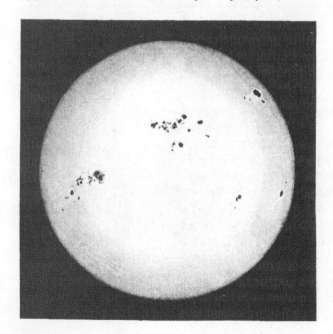

Photograph showing high level of sunspot activity in 1870

Observations of the corona over the following decades proved this theory correct. An eclipse that occurs near a low point in the sunspot cycle (as in 1878) reveals a corona that is dimmer than normal but that shows a more detailed structure. The "halo" seen behind the black disk of the Moon is somewhat compressed. The long *equatorial streamers* may be seen stretching out from either side of the Sun. Finely detailed *polar plumes* of light curve above and below the dark disk in the sky. At sunspot maximum, the appearance is just the opposite. The plumes and streamers are less pronounced, but the coronal glow around the Sun is brighter and more expanded. Eclipses occurring at intermediate stages of the cycle (as in 1979) exhibit some combination of features of both types.

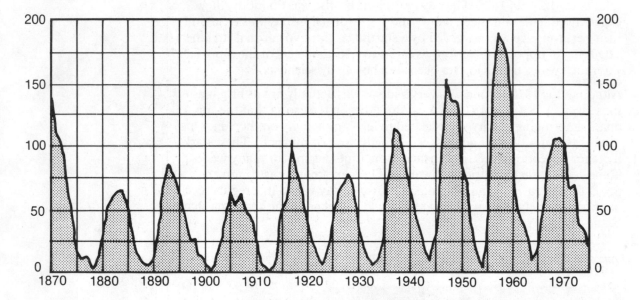

Graph of yearly sunspot numbers

The 1878 eclipse also marked the height of the search for the elusive planet "Vulcan." This was the name given to a small object reportedly seen near the Sun on several occasions; some scientists thought it was a planet that up to then had escaped detection. The calculation of small irregularities in the orbit of Mercury supported this theory. Just 30 years earlier Neptune had been discovered in a similar manner. Astronomers were hoping that the clear skies and the blacked-out Sun would reveal the planet in their telescopes during the eclipse. But it didn't happen. One astronomer did announce he had discovered Vulcan, but he was later proved wrong. The discrepancies in Mercury's orbit are fully accounted for by Einstein's theory of relativity.

Because the Sun's light is shielded during an eclipse, some of the brighter stars and planets can be seen in the darkened sky. This fact has enabled astronomers to test part of the theory of relativity. According to Einstein, who proposed the theory in 1915, rays of light should be deflected by a gravitational field. In particular, starlight passing near the Sun should be bent slightly toward the Sun. The only time when stars near the Sun are visible is during a total solar eclipse.

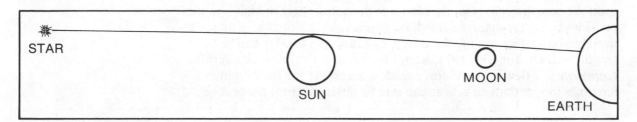

Bending of starlight during an eclipse

Scientists put the theory to test during eclipse expeditions to Brazil in 1919 and to Australia in 1922. Photographs of stars near the Sun during these eclipses were compared to photographs taken of the same stars several months later when the Sun, in another part of the sky, would have no effect. The difference in position of the stars showed that Einstein was correct. Eclipses became the first tool to crack the door of experimental proof on one of the most profound scientific ideas about the universe.

Scientists at these "relativity eclipses" were blessed with clear skies. The excitement of this new theory led to great plans for observation of totality on September 10, 1923. The path would graze Southern California at a favorable time of year. Weathermen predicted a 90% chance of clear skies. But as fate would have it, the day was overcast and the clouds spoiled all the planned observations.

A year and a half later there was little hope of good weather for the total eclipse in the Northeast United States. Yet many places in the path of totality in New York and Connecticut experienced clear skies on January 24, 1925. Millions of people witnessed the eclipse. The southern edge of the path crossed right through New York City. This situation provided a unique opportunity to determine the precise location of the edge of totality during an eclipse.

Astronomers knew beforehand that the edge of the Moon's shadow would cut Riverside Drive in New York City somewhere between 83rd and 110th Streets. To be on the safe side, observers were positioned at every intersection between 72nd and 135th Streets. They were instructed to report whether they had seen the corona (total phase) or only a crescent of the Sun (partial phase). The results were definite: the edge of the umbra passed between 95th and 97th Streets, yielding an accuracy of several hundred feet for a shadow cast a distance of over 200,000 miles.

In recent years, great technological advances have occurred in many types of instruments used to gather scientific data during eclipses. Much of this increased sensitivity is lost, however, because of distortion by the lower atmosphere. More accurate data can be gathered at higher elevations, but eclipse paths don't often pass over convenient mountaintops. The modern astronomer's solution has been to take to the air. Three distinct advantages have resulted from the use of "flying observatories" in the past few decades. First, flying above the clouds ensures that bad weather will not spoil the occasion. Second, the clarity of the atmosphere at high altitudes provides better results. And third, because an aircraft can fly in the direction the shadow is moving, the effective duration of totality can be lengthened. On June 30, 1973, scientists aboard the supersonic aircraft *Concorde 001* flew in the Moon's shadow across Africa for 74 minutes — ten times longer than an eclipse can ever be observed from the ground.

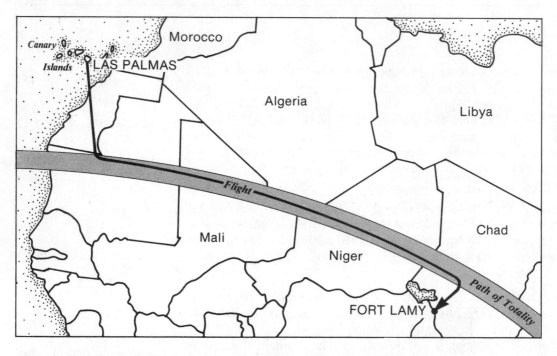

Flight path of Concorde 001 *in the Moon's shadow on June 30, 1973*

But the story doesn't end there. The study of the Sun is also reaching into space. Observations of the solar corona from the Skylab orbiting space station have expanded our understanding of the Sun into areas previously unexplored. This increase in solar knowledge coincides with a growing public awareness of the Sun as the primal source of energy for our planet. A total solar eclipse provides a magnificent opportunity to personally appreciate this source of life-giving energy at the center of our solar system. The next part of this book explains what happens during an eclipse and how these events are repeated in time and space.

Chapter 2

Understanding Eclipses

The Approach of Darkness

A total eclipse begins almost unnoticeably. First contact occurs when the Moon starts its passage across the face of the Sun. At first, only a small "bite" appears on the western edge of the Sun. Gradually, as more and more of the Sun disappears, an interesting shadow effect can be seen. The tiny spots of light shining through the leaves of a tree, for example, show up on the ground as crescent images of the slowly vanishing Sun.

This partial phase of the eclipse leads to totality in about an hour, sometimes slightly longer. For most of that time, there is little hint of the approaching darkness. But as the bright area of the Sun is reduced more and more, the increasing darkness becomes noticeable. Daylight fades very quickly in the last few minutes before totality.

Crescent images of partially eclipsed Sun (1900 engraving)

Shadow bands of a solar eclipse (1900 engraving)

While a small crescent of the Sun remains in the sky, a curious eclipse phenomenon is often observed. Thin wavy lines of alternating light and dark can be seen moving and undulating in parallel on plain light-colored surfaces. These so-called *shadow bands* are the result of sunlight being distorted by irregularities in the Earth's atmosphere. An open floor or wall is a good place to look for them. A similar effect is seen when the Sun shines through ripples on the surface of the water in a swimming pool; the wavy lines moving on the bottom of the pool resemble the shadow bands of an eclipse.

As the narrow crescent of the Sun finally begins to disappear, tiny specks of light remain visible for a few seconds more. These points of light are spaced irregularly around the disappearing edge of the Sun, forming the appearance of a string of beads around the dark disk of the Moon. These lights are known as *Baily's beads*, named after Francis Baily, the 18th century English amateur astronomer who was the first to draw attention to them.

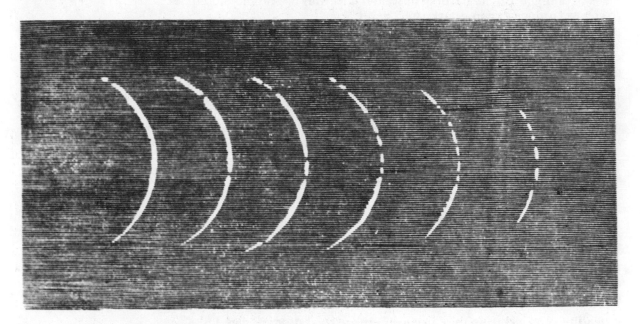

Baily's beads (engraving from the eclipse of July 18, 1860)

Baily's beads would not be possible if the Moon's surface were perfectly smooth. The edge of the Sun is first hidden by the peaks of lunar mountains. The beads are the last few rays of sunlight shining through valleys on the edge of the Moon. Baily's beads make their brief appearance up to 15 seconds before totality. When a single point of sunlight remains, a beautiful "diamond ring" effect is created against the outline of the Moon. This final sparkling instant signals the climax of the eclipse. The last ray of sunlight vanishes and totality begins.

> *O dark, dark, dark, amid the blaze of noon,*
> *Irrevocably dark, total eclipse*
> *Without all hope of day.*
>
> *Milton,* Paradise Regained

The Spectacle of Totality

Suddenly the sky above is dark. The Moon's shadow, racing along the Earth at speeds up to several thousand miles per hour, brings a swift and dramatic nighttime effect. The sky near the horizon, where the eclipse is not total, still appears bright. This distant scattered light produces a slight reddish glow and unusual shadow effects. This ''darkness'' is not quite as black as at night. But its startling onset and unearthly appearance combine to create a unique daytime darkness.

In the center of this darkened sky hangs the featured spectacle of the eclipse — the corona of the Sun. This pearly white crown of light shines in all directions around the darkened solar disk. A million times fainter than the Sun itself, the full glory of the corona is visible only during a total solar eclipse.

The corona consists of the ionized gases that form the outer atmosphere of the Sun. Although these gases extend many millions of miles into space, only the corona near the Sun is visible to the naked eye. Wispy plumes and streamers of coronal light reach out distances up to several diameters of the Sun before they fade into darkness.

The corona comes into full view when the leading edge of the Moon blots out the Sun, and it remains visible throughout totality. For a few seconds both after the beginning and before the end of totality, a pinkish glow appears at the edge of the Moon. This glow is light from the Sun's lower atmosphere, the *chromosphere*. Its rosy color (''chromo'' means color) comes from its main element, hydrogen.

Extending outward from the chromosphere are *solar prominences*. Usually several of these red cloudlike formations are visible during a total eclipse. Some prominences actually erupt, speeding away from the Sun at close to a million miles per hour. They arch above the surface and then disappear, sometimes lasting only a matter of hours. A few of these erupting prominences have been seen to reach a height of nearly one-third the diameter of the Sun itself.

This marvelous view of the Sun clearly commands the center of attention during totality. But there are other sights to see as well. Because the direct light of the Sun is blocked, some of the brighter stars and planets become visible. Sometimes a total solar eclipse reveals a small comet on its path near the Sun.

The darkness of totality resembles nighttime, and plants and animals react accordingly. Birds stop singing and may go to roost. Daytime flower blossoms begin to close as if for the night, and bees get disoriented and stop flying. The temperature drops in the coolness of the Moon's shadow. All of Nature seems still and quiet for this brief moment of daytime darkness.

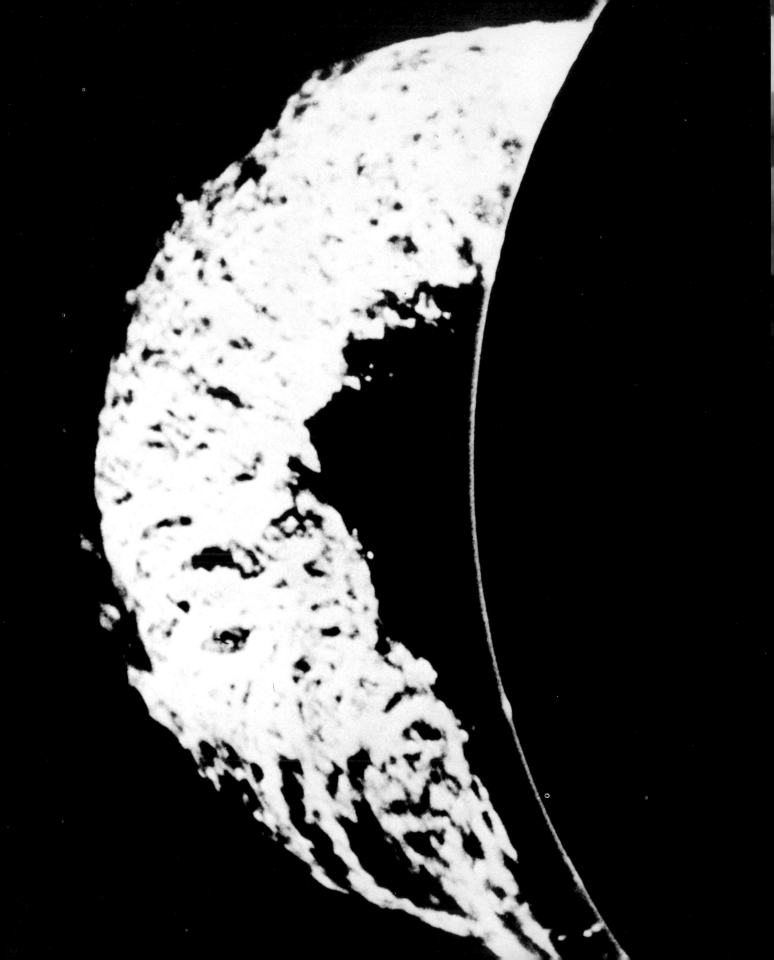

And then the shadow passes. A bright speck of sunlight flashes into view at the eastern edge of the Sun as the corona disappears. Totality has ended. The same events that preceded totality now occur in reverse order and on the opposite side of the Sun. Baily's beads appear, followed by a thin crescent of the Sun. Daylight returns as more and more of the Sun is gradually uncovered by the passing Moon.

Finally the complete disk of the Sun is restored. The eclipse is over. The Moon continues in its orbit around the Earth, casting its shadow off into the vastness of space. Nothing tangible remains of the eclipse except some photographs and scientific data. Yet the memory of the experience is permanent — the fleeting beauty of the corona etched into the mind's eye by the sheer grandeur of the event. There is simply nothing else like it. And now it is gone — but not forever. The patterns of time and space will repeat themselves to create other solar eclipses.

Engravings from the eclipse of August 7, 1869

Page 37: Diamond-ring effect at eclipse of October 2, 1978
Page 38: Eclipse of June 8, 1918, Goldendale, Washington
Page 39: Corona and Venus during eclipse of November 12, 1966, Bolivia
Opposite: Giant solar prominence photographed on June 4, 1946

Patterns in Time and Space

In the 20th century, only nine total eclipse tracks cross the United States (excluding Alaska and Hawaii). The longest path of totality across the country came with the eclipse of June 8, 1918. This so-called "American eclipse" was observed from one corner of the nation to the other, all the way from the state of Washington to Florida. Astronomers were joined by crowds of interested people along the eclipse track to witness the wonder of this daytime darkness. Newspaper headlines of World War I shared the day with news of the Sun's eclipse.

The final total eclipse visible from the United States in this century is the eclipse of February 26, 1979. The next one after that won't be until August 21, 2017. Of course, there will be total eclipses in between those dates; but none of them can be seen from any of the contiguous 48 states.

The sequence of eclipses from year to year is determined by two different cycles of the Moon. The familiar monthly change of the phases of the Moon is one of these. The other cycle involves the gradual shift in orientation of the Moon's orbit. Only when these two cycles are favorably combined (about every six months) can a solar eclipse occur.

A solar eclipse must take place at a new Moon. During this lunar phase the Moon passes between the Earth and the Sun. The Sun shines on the side of the Moon facing away from us, casting a shadow toward the Earth. A new Moon appears every 29½ days, but usually the Moon's shadow passes completely above or completely below the Earth. This is because the Moon's orbit is tilted at a slight angle to the Earth's orbit; the Moon usually passes above or below the direct line of sight between the Earth and the Sun.

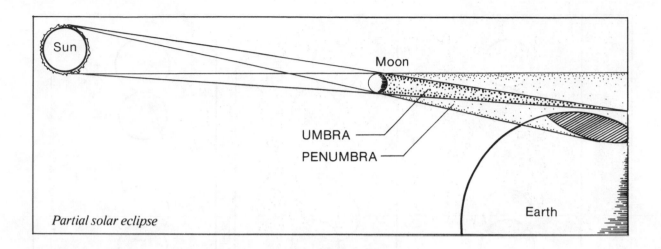

Partial solar eclipse

A total eclipse occurs when the umbra (complete shadow of the Moon) sweeps across the Earth. During a *partial eclipse*, only the penumbra (partial shadow of the Moon) touches our planet. The umbra passes either just above the North Pole or just below the South Pole, completely missing the Earth. No total eclipse is visible — only partial phases can be seen. It has the same appearance as the partial phases of a total eclipse, but is visible only near the North or South Pole.

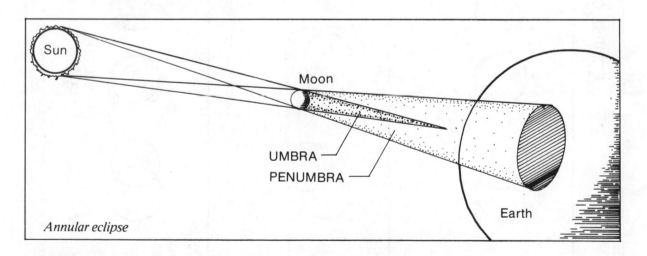

Annular eclipse

A third type of solar eclipse occurs when the Moon's umbra passes across the Earth, but is not quite long enough to touch the surface; the shadow cone diminishes to a point before reaching the Earth. This effect happens when the Moon is farther out in its orbit around the Earth. The Moon appears slightly smaller and is not large enough to completely cover the Sun. When the Moon is centered over the Sun, a ring of sunlight remains visible around the edge. This type of eclipse is called an *annular eclipse*. (Annular comes from the Latin word meaning ''ring.'')

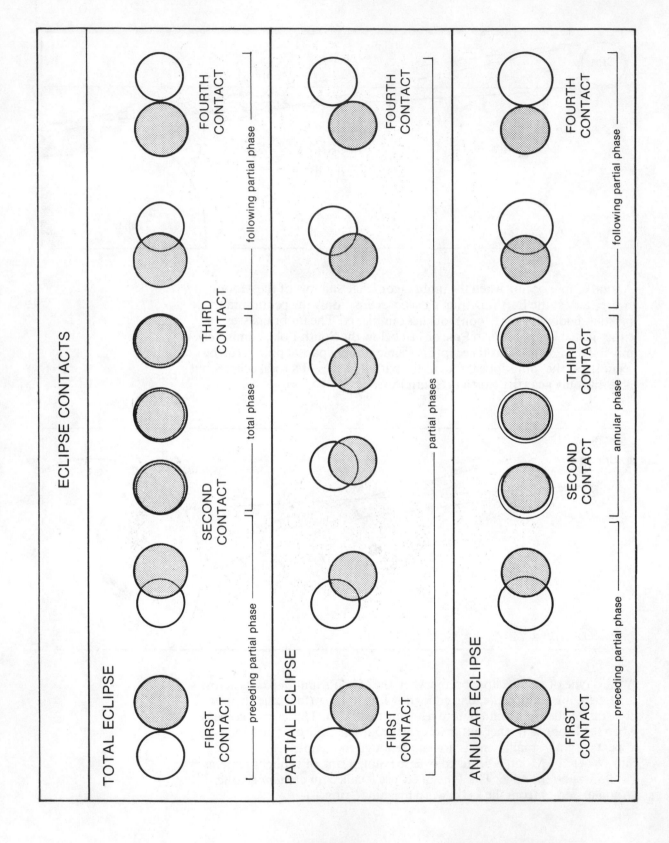

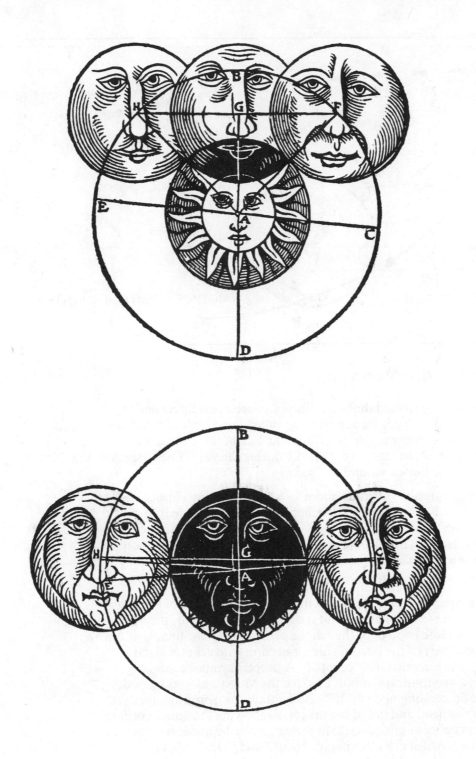

Diagrams of contacts for partial eclipse (top) and total eclipse (bottom) from Peurbach's Theorica Novae Planetarum *(1553)*

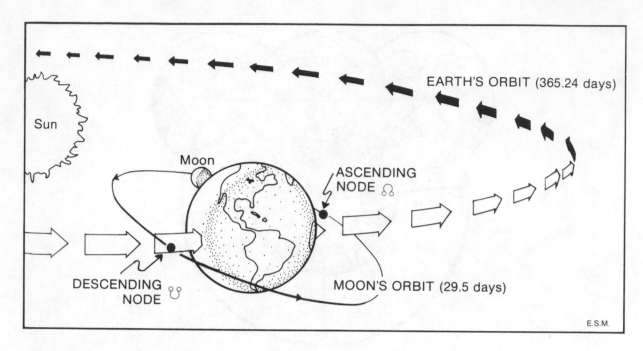

Diagram labels: Sun — Moon — EARTH'S ORBIT (365.24 days) — ASCENDING NODE ☊ — DESCENDING NODE ☋ — MOON'S ORBIT (29.5 days) — E.S.M.

The Orbit of the Moon

As the Earth moves around the Sun in its yearly orbit, the direct line of sight between them sweeps through a plane called the *ecliptic*. The orbit of the Earth defines this plane. (As seen from the Earth, the ecliptic is the path the Sun appears to take across the sky during the year. The 12 signs of the zodiac are distributed around this path.)

The orbit of the Moon is tilted by about five degrees to the ecliptic. Half of the time the Moon will be above the ecliptic plane, the other half of the time below. The two points where the Moon's orbit intersects the ecliptic are called the *nodes*. The *ascending node* marks the point of the Moon's passage to the upper part of its orbit; the *descending node* is the point where the Moon moves into its orbit below the ecliptic plane.

For a solar eclipse to occur, the Moon must be at or near one of its nodes when it passes between the Earth and the Sun. In other words, the new Moon must be close enough to the ecliptic plane so that the lunar shadow will cross some part of the Earth. This connection with eclipses is the reason the plane is named the "ecliptic." A deeper, symbolic meaning is found in the astronomical symbols used for the Moon's ascending node (☊) and the descending node (☋). These symbols are generally supposed to represent the head and tail of the dragon swallowing the Sun according to the ancient belief about eclipses. In earlier times, the nodes were actually known by the fanciful titles "Dragon's Head" and "Dragon's Tail."

Opposite: Petrus Apianus' Astronomicum Caesareum *(1540) includes a rotating* volvelle *used to compute the position of the lunar nodes (depicted as the head and tail of a dragon).*

46

F III

47

Corona seen on December 22, 1870

*Methinks it should be now a huge eclipse
Of Sun and Moon.*

Shakespeare, Othello

Eclipse Seasons

The Moon and its orbit naturally move with the Earth as it travels around the Sun every year. If the lunar nodes were stationary (with respect to the stars), the ascending node would be lined up between the Earth and the Sun at the same time each year (likewise for the descending node half a year later on the opposite side of the Earth's orbit). But the nodes of the lunar orbit are not quite stationary; they are gradually shifting their orientation in space. By the time a node is in line with the Sun again, it has advanced slightly. The alignment happens sooner than if the nodes were not moving. Thus it takes less than a full year for a node to be realigned between the Earth and the Sun. This period, called the *eclipse year*, is about 346.6 days long.

The diagram on the opposite page illustrates the progression of the Moon's nodes through an eclipse year. At position A the ascending node is lined up between the Earth and the Sun. This is the beginning of an eclipse year. As the Earth moves on (as in position B), the node passes out of the Earth-Sun alignment. About six months later position C is reached where the descending node lines up. But because the nodes themselves are slowly advancing, this alignment occurs a few days before the Earth reaches the point exactly opposite A. As the year continues, the descending node passes out of alignment (as in position D). Finally, at position E, the ascending node returns to a position between the Sun and the Earth. But because of the slow revolution of the nodes, the return takes only 346.6 days, an eclipse year. This is 18.6 days short of the full year it takes for the Earth to return to A.

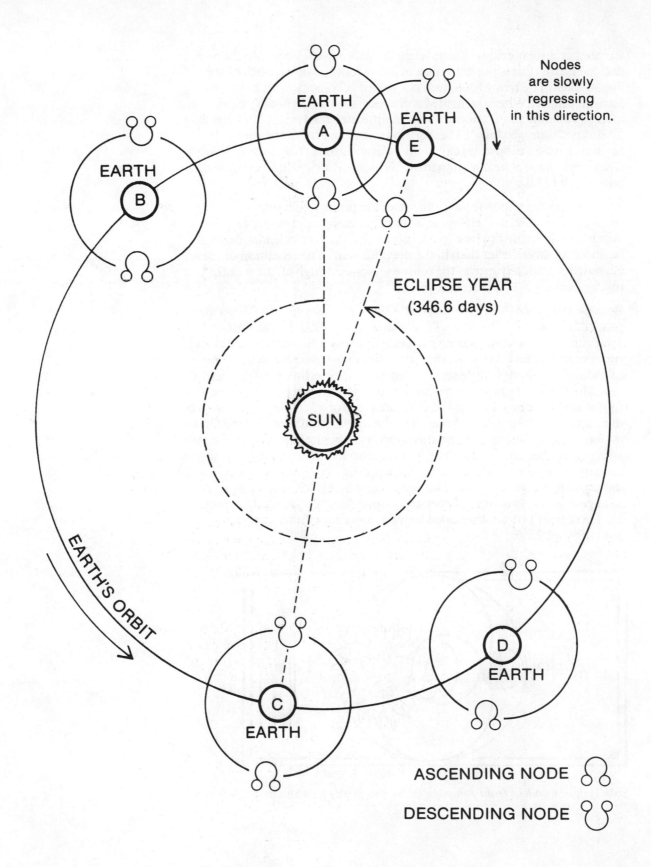

EARTH A

EARTH E

EARTH B

Nodes are slowly regressing in this direction.

ECLIPSE YEAR
(346.6 days)

SUN

EARTH'S ORBIT

D
EARTH

C
EARTH

ASCENDING NODE

DESCENDING NODE

Recall that a solar eclipse occurs when the Moon passes between the Sun and the Earth (this is the definition of new Moon) at or near one of the lunar nodes. The new Moon need not perfectly coincide with a node to have an eclipse. When the new Moon appears within 18¾ days before or after the alignment of a node, a solar eclipse may take place. This creates a 37½-day "time window." These periods when the conditions are favorable for an eclipse are called the *eclipse seasons*. These seasons occur whenever a node is near alignment. This happens twice each eclipse year, once every 173 days.

The table on the opposite page illustrates the timing of all solar eclipses from 1954 to 1979. The eclipse seasons occur earlier and earlier each year. When a node returns to its alignment with the Sun, the calendar date will be about 18 days earlier than in the previous year. The repetition of these alignments gradually moves the eclipse seasons through all the months in a regular pattern over a period of years.

Because the new Moon cycle lasts 29½ days, at least one new Moon appears during every 37½-day eclipse season. As it turns out, at least one solar eclipse must occur during each eclipse season. Since there are at least two eclipse seasons every year, there must be at least two solar eclipses every calendar year. Some eclipse seasons have two solar eclipses. In this case a new Moon appears near the beginning and the following one near the end of the time window. Two of these "double seasons" in the same year produce a total of four solar eclipses. The maximum number of solar eclipses in one calendar year is five and this happens only rarely. In this case a year with two double seasons has the fifth eclipse squeezed in at the beginning of January or the end of December. This is possible because the eclipse year is shorter than a calendar year. This last occurred in 1935 when there were solar eclipses on January 5, February 3, June 30, July 30, and December 25. There won't be another calendar year containing five solar eclipses until the year 2206.

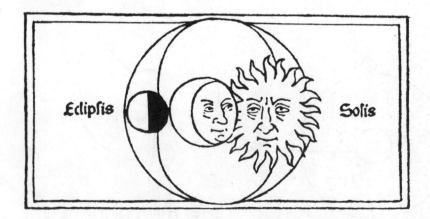

Solar eclipse woodcut from Johannes de Sacrobusco's Opus Sphaericum *(1482)*

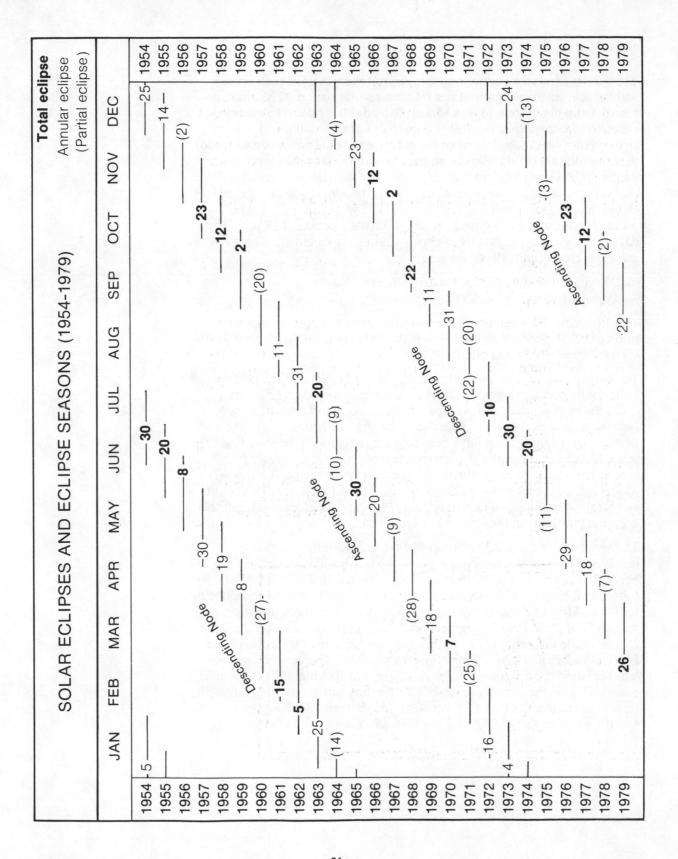

SOLAR ECLIPSES AND ECLIPSE SEASONS (1954-1979)

Total eclipse

Annular eclipse

(Partial eclipse)

The *Saros* Cycle

The eclipse seasons repeat year after year, but the timing of each eclipse within successive seasons does not follow a regular pattern. The lunar month from new Moon to new Moon (29.53 days) is called the *synodic month* (from the Greek word for "meeting" or "conjunction"). An eclipse year (346.62 days) does not come close to being an exact multiple of these periods (324½ days in eleven synodic months, 354 in twelve). A longer cycle, close to an exact multiple of these two periods, would be useful for making eclipse predictions. This is the *saros* cycle discovered by Babylonian astronomers in ancient times. The *saros* (meaning "repetition") lasts exactly 223 synodic months. That's a period of 18 years 11 days (or 18 years 10 days if five February 29th's are included). The *saros* coincides closely with 19 eclipse years:

> 223 synodic months (29.5306 days) = 6,585.32 days
> 19 eclipse years   (346.6200 days) = 6,585.78 days

This resonance between the periods of these two cycles produces a repetition of eclipses in a remarkably short time. (In terms of astronomical cycles, 18 years is a short time!)

To illustrate how the *saros* works, see the table of eclipses from 1954 to 1979 on the previous page. A total solar eclipse occurred on February 15, 1961, the path of totality crossing southern Europe and parts of Russia. Because a solar eclipse took place, we know that the Moon must have been new and that a node must have been near alignment with the Sun. Eighteen years and eleven days later the *saros* cycle repeats. Because nearly 19 eclipse years have gone by, the same node is near alignment. It is a new Moon again because exactly 223 synodic months have passed. This results in a total solar eclipse on February 26, 1979. The cycle is repeated, but this eclipse is visible from the United States and Canada.

The *saros* has an extra 0.32 portion of a day included in its period (6,585.32 days). When the cycle repeats, the Earth will have rotated beyond its position at the former eclipse by this fraction of a day. The subsequent eclipse will be seen about a third of the way around the globe to the west. After three *saros* cycles, an eclipse takes place near the original longitude of the eclipse 54 years earlier. However, each eclipse in the series moves a little farther in the same direction toward one of the poles of the Earth. The February 26, 1979, eclipse track falls at slightly more northerly latitudes than the February 15, 1961, eclipse. This gradual shift in latitude occurs because the new Moon at each succeeding *saros* moves slightly wtih respect to the node. The half-day difference between 19 eclipse years and the *saros* (6,585.78 - 6,585.32 = 0.46 days) causes this change.

A series of eclipses each separated by this 18-year 11⅓-day cycle is called a *saros series*. Because the resonance between 19 eclipse years and the *saros* is not exact (0.46-day difference), a *saros* series cannot go on indefinitely. Eventually a series reaches a point when the eclipses are no longer visible; the umbra passes too far above or below the Earth to be seen. A *saros* series spans over 1,200 years and includes between 68 and 75 solar eclipses. The *saros* series that includes the February 26, 1979, eclipse will end in the year 2195.

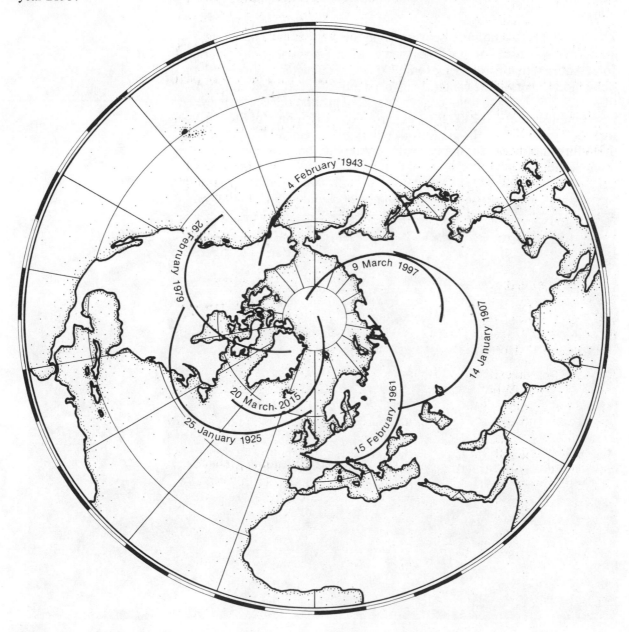

Polar projection map showing several eclipse tracks from same saros *series*

The Secret of Stonehenge

There is no evidence that the builders of Stonehenge knew of the *saros*. But they didn't need to. The *saros* is a coincidence of celestial cycles; the marking of the monthly and yearly cycles at Stonehenge would work fine even if the *saros* did not exist. The likely purpose behind the pattern of these ancient stones and markers — to predict eclipses — has been revealed by the studies of two modern astronomers, Gerald Hawkins and Fred Hoyle.

The key to this fascinating discovery lies in the number 56. That is the number of so-called Aubrey holes (named after 17th century antiquarian John Aubrey) that form a ring some 300 feet in diameter around the large stone monuments in the center. Dug to a depth of several feet and then filled with chalk, these holes are evenly spaced around the circle. They were established about 2400 B.C. as part of the early phase of Stonehenge, some 300 years before the giant stone pillars were erected. The purpose of the Aubrey holes was for many years a mystery to those who studied Stonehenge.

Light-colored Aubrey holes (more evident at left) form a circle around Stonehenge.

54

In his 1965 book *Stonehenge Decoded*, astronomer Gerald Hawkins explained his interpretation of the Aubrey holes. How does the number 56 relate to eclipses? Recall that the nodes of the Moon's orbit are regressing; they are moving slowly around the Earth from east to west. A complete revolution of these invisible orbital points takes 18.6 years (not to be confused with the 18-year *saros* cycle). To accurately predict eclipses from year to year, this regression cycle must be carefully recorded. Yet the builders of Stonehenge did not (as far as we know) use any form of writing. How could they mark this cycle? Three times 18.6 is very nearly 56. According to Hawkins, moving a marker such as a stone three Aubrey holes each year makes a complete revolution of the system in 18⅔ years, close enough to the actual value for use in eclipse predictions. Two of these markers, on opposite sides of the Aubrey ring, would show the position of the nodes at all times.

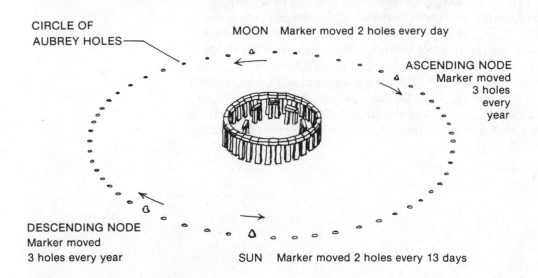

CIRCLE OF
AUBREY HOLES

MOON Marker moved 2 holes every day

ASCENDING NODE
Marker moved
3 holes
every
year

DESCENDING NODE
Marker moved
3 holes every year

SUN Marker moved 2 holes every 13 days

There is no direct historical record to confirm or refute this idea of Stonehenge as a primitive "computer" of eclipses. The ancient monument itself and its alignment with the Sun and the Moon are all we have to go on. Yet much can be understood from this information. It seems clear that Stonehenge was used to mark the limits of the rising and setting of the Sun and the Moon. These astronomical alignments, too precise to be mere chance, also imply a knowledge of the 18.6-year cycle of the lunar nodes. It seems plausible that the Aubrey holes were used to mark this cycle. What's more, the number 56 may also be used to mark the movement of the Sun and the Moon. According to Hoyle, if a Sun marker is moved two holes every thirteen days, a complete circuit takes 364 days (56/2 x 13); this is only 1¼ days short of a full year. If a Moon marker is moved two holes every day, a complete circuit (28 days) is very close to the lunar month. These small discrepancies are no bother because they can be easily corrected by monthly and yearly observations.

Viewed in this way, Stonehenge represents a working model of the Sun-Moon-Earth system. The Sun, the Moon, and the lunar nodes, each represented by markers, revolve around the Earth located in the center. When the markers coincided, an eclipse would take place. All the necessary information (periodically corrected by actual observation) would have been available to the Stonehenge "astronomers" to use the 56 Aubrey holes as a sort of primitive computer to predict eclipses.

But the story may not end there. A further refinement of eclipse calculations seems possible. Recall that a solar eclipse may occur if there is a new Moon within a certain period either before or after the alignment of a node. (These are the eclipse seasons explained earlier in this chapter.) That period is ± 18.8 days. These time windows could be counted using the 56 holes by moving a marker three holes every day two times around the circle. But to use the Aubrey holes to accurately mark these eclipse seasons would require a precise knowledge of when the nodes are in alignment. This is the cycle of the eclipse year. The people of Stonehenge knew the length of the full year (365¼ days). This is derived by counting the days between successive summer solstice sunrises. How might they have kept track of the shorter eclipse year (346.6 days)? The difference between the two years happens to be 18.6 days. Here again a period of 18.6 days could be counted using the 56 holes. Thus the Aubrey holes could serve a triple purpose of marking the regression of the lunar nodes (18.6-year cycle), the eclipse seasons (± 18.8-day time window), and the eclipse year (18.6-day difference from full year).

Opposite: A Mythological Trilithon, from William Blake's Jerusalem. *The oversized stones frame three figures (Bacon, Newton, and Locke) standing below what appears to be the partially eclipsed Moon. Blake created this engraving in the early 19th century, long before anyone reasoned a connection between Stonehenge and eclipses.*

And this the form of mighty Hand sitting on Albions cliffs
Before the face of Albion: a mighty threatning Form.

His bosom wide & shoulders huge overspreading wondrous
Bear Three strong sinewy Necks & Three awful & terrible Heads
Three Brains in contradictory council brooding incessantly.
Neither daring to put in act its councils, fearing each other.
Therefore rejecting Ideas as nothing & holding all Wisdom
To consist. in the agreements & disagreents of Ideas.
Plotting to devour Albions Body of Humanity & Love.

Such Form the aggregate of the Twelve Sons of Albion took; & such
Their appearance when combind: but often by birth-pangs & loud groans
They divide to Twelve: the key-bones & the chest dividing in pain
Disclose a hideous orifice; thence issuing the Giant-brood
Arise as the smoke of the furnace. shaking the rocks from sea to sea.
And there they combine into Three Forms, named Bacon & Newton & Locke,
In the Oak Groves of Albion which overspread all the Earth.

Imputing Sin & Righteousness to Individuals; Rahab
Sat deep within him hid: his Feminine Power unreveald
Brooding Abstract Philosophy. to destroy Imagination. the Divine-
Humanity A Three-fold Wonder: feminine: most beautiful: Three-fold
Each within other. On her white marble & even Neck, her Heart
Inorbd and bonified: with locks of shadowing modesty, shining
Over her beautiful Female features, soft flourishing in beauty
Beams mild, all love and all perfection, that when the lips
Recieve a kiss from Gods or Men, a threefold kiss returns
From the pressd loveliness: so her whole immortal form three-fold
Three-fold embrace returns: consuming lives of Gods & Men
In fires of beauty melting them as gold & silver in the furnace
Her Brain enlabyrinths the whole heaven of her bosom & loins
To put in act what her Heart wills; O who can withstand her power
Her name is Vala in Eternity: in Time her name is Rahab

The Starry Heavens all were fled from the mighty limbs of Albion His

Motion of the Moon's Shadow

The builders of Stonehenge created a monument that embodies all the information needed to predict eclipses, including a way of keeping track of the nodes of the lunar orbit. These invisible points in space — symbolized as dragons or serpents by ancient cultures — help determine the path of the Moon's shadow as it moves across the Earth toward the east during an eclipse.

Why does the shadow move eastward? Both the Moon and the Sun "rise" in the east and "set" in the west. This apparent motion across the sky (from east to west) is the result of the daily rotation of the Earth. Our planet is steadily spinning eastward; thus objects in the heavens (including the Sun and the Moon during an eclipse) seem to move toward the west. For a place in the path of totality, the entire duration of an eclipse (from first contact until fourth contact) is about two hours. During that time the Sun and Moon move through a part of their path across the sky from east to west.

The Moon's shadow, however, moves in the opposite direction: eastward. This happens because the Moon is revolving in its orbit from west to east. The umbra moves eastward with the Moon as it passes between the Sun and the Earth. This creates the effect of the Sun seeming to overtake the Moon during an eclipse. Because the Sun is farther away, it passes westward behind the Moon and casts a shadow that moves eastward on the Earth.

As the umbra sweeps eastward, an observer located in the path is also moving eastward due to the rotation of the Earth. But the Moon's shadow moves faster than any point on our rotating planet; the umbra always overtakes a stationary observer in its path. The faster the shadow is moving, the shorter is the duration of totality. How fast the umbra moves over the Earth's surface depends mainly on two factors: (1) the latitude (distance from the equator) of the point in the path, and (2) the time of day totality occurs for that point.

The first factor is the latitude. A point on the equator travels the complete circumference of the Earth (nearly 25,000 miles) in twenty-four hours, a rate of about 1,040 miles per hour. Points at higher latitudes (either farther north or farther south) don't have as far to go around the Earth in a day's rotation. The speed of a point on the Earth becomes progressively slower at greater and greater distances from the equator. But the umbra's speed through space in the vicinity of the Earth is the same (about 2,100 miles per hour) regardless of latitude. The difference between this value and the speed of the point on the Earth is the speed of the umbra across that point. The result is that the farther away from the equator, the larger is this difference, and thus the faster is the shadow. Points farther from the equator move slower and the umbra overtakes them faster.

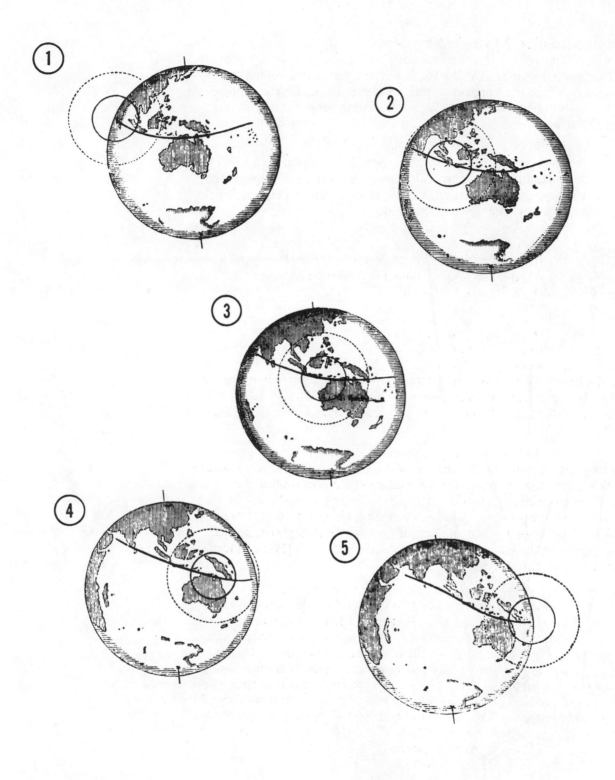

Motion of the Moon's shadow during the eclipse of December 12, 1871. Notice how the Earth rotated about 45 degrees from the start to the finish of the eclipse.

The other major factor affecting the speed of the shadow is the time of day the eclipse takes place. The umbra moves slowest across places where totality happens at noontime. When totality occurs earlier or later in the day, the umbra strikes the Earth at an oblique angle. The shadow at sunrise or sunset moves faster than if it were closer to being perpendicular to the Earth's surface.

The combination of these two factors — latitude and time of day — determines the speed of the umbra. The speed is important because it affects the duration of the eclipse: the slower the shadow, the longer the time of totality. But another factor is just as important in determining the duration of totality: the width of the shadow.

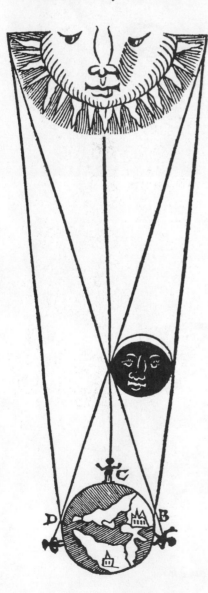

Solar eclipse from Peurbach's Theoricae Novae Planetarum *(1553)*

Opposite: James Ferguson's "Eclipsareon" from Chamber's Encyclopaedia *(London, 1779). This astronomical contrivance could exhibit the "time, quantity, duration, and progress of solar eclipses." The date and time of an eclipse would first be set on the dials at the base of the globe. Then the frame containing the screen of concentric circles would be adjusted for the Moon's latitude. A light source (such as a candle) would be used to project the shadow. By turning the crank, the circular screen would move across the frame, casting its simulated umbra on the rotating globe.*

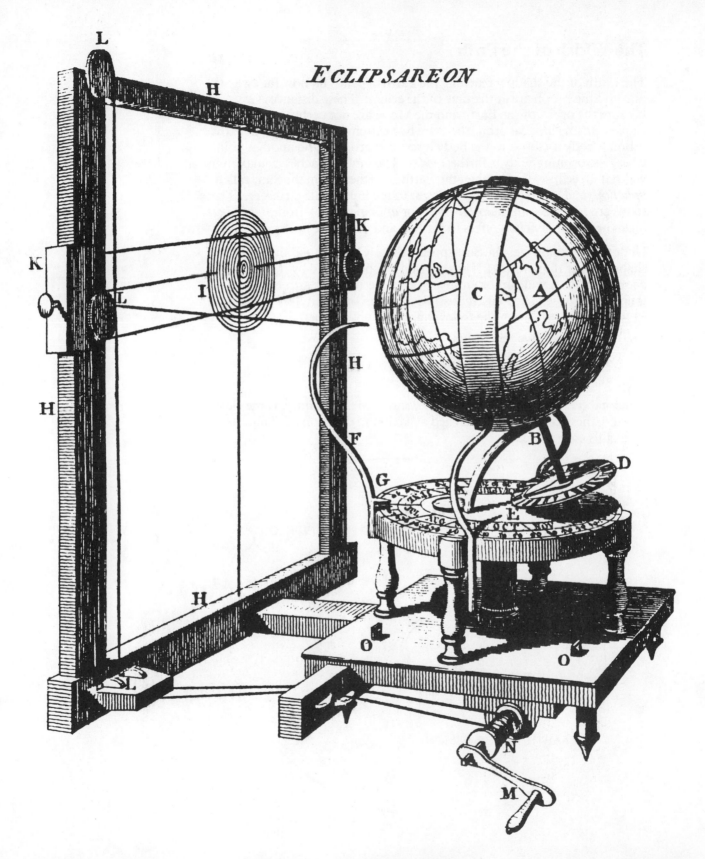

ECLIPSAREON

The Width of the Path

The width of the shadow reaching the Earth depends on how far away the Sun and the Moon are at the time of the eclipse. These distances vary because the orbits of the Earth and the Moon are not perfect circles, but ellipses. In an elliptical orbit, there are two extreme points: one where the orbiting body is closest to the body it revolves around, and another point where the orbiting body is farthest away. The umbra reaches its maximum width if an eclipse occurs when the Earth is farthest from the Sun, called *aphelion*, and when the Moon is closest to the Earth, called *perigee*. (These terms are formed with the prefixes *ap-* or *apo-* meaning "from," and *peri-* meaning "near"; *helion* refers to the Sun and *gee* to the Earth.)

Under these conditions the Sun appears slightly smaller and the Moon slightly larger than normal. Thus the Sun can be covered by the Moon for a longer time. A wide umbra moving slowly across the Earth produces a long total eclipse. The longest possible duration of totality, 7 minutes and 31 seconds, occurs when the following conditions are met:

(1) The observer is at the equator;
(2) Totality occurs there at noon;
(3) The Earth is farthest from the Sun (aphelion); and
(4) The Moon is closest to the Earth (perigee).

The total eclipse of June 30, 1973, the longest of this century, came close to meeting these conditions: it reached a maximum duration of 7 minutes and 3 seconds over Africa.

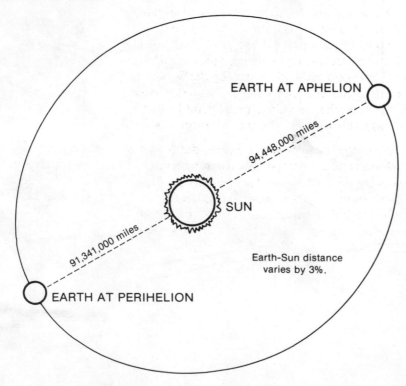

EARTH AT APHELION

94,448,000 miles

SUN

91,341,000 miles

Earth-Sun distance
varies by 3%.

EARTH AT PERIHELION

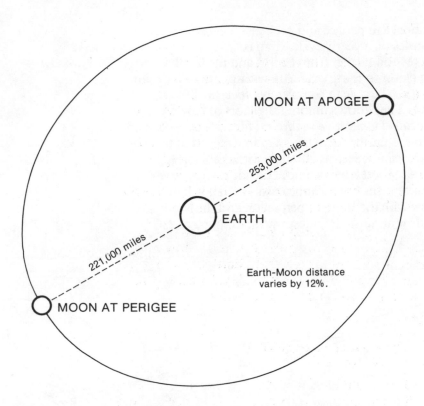

MOON AT APOGEE

253,000 miles

EARTH

221,000 miles

Earth-Moon distance
varies by 12%.

MOON AT PERIGEE

The distances from the Earth to the Sun and to the Moon vary in regular cycles. The Earth orbits the Sun once a year, reaching its closest point (perihelion) in early January and its farthest point (aphelion) in early July. However, these orbital points are not stationary; each year they occur a fraction of a day later in time. Their slow revolution around the Sun takes about 20,000 years for a complete cycle. Ten thousand years ago the positions were reversed: perihelion was reached in July and aphelion in January. Ten thousand years from now this will again be the situation.

The Moon's orbit follows a similar motion around the Earth, but the time scale is greatly reduced. The Moon returns to its perigee (or apogee) in about two days less than it takes for a full Moon cycle. This shorter 27½-day cycle is called the *anomalistic month*. This is the time it takes for the Moon to return to its original position the same distance from the Earth. These orbital points are slowly revolving around the Earth, making a complete revolution every 8.85 years.

The Moon passes from apogee to perigee and back again every 27½ days; the Earth-Sun distance varies on a yearly cycle. If an eclipse occurs when the Moon is near apogee (Moon farther from Earth) and the Earth is nearer to perihelion (Sun closer to Earth), the Sun will appear larger than the Moon. The Moon will not be able to completely block the Sun; the result is an annular eclipse. The Moon's umbra falls short of reaching the Earth, producing what is called a *negative shadow*. This is an extension of the umbra projected onto the Earth. Anywhere within the path of this negative shadow the eclipse can be seen as annular, with the Sun completely surrounding the Moon from behind. Outside the negative shadow, within the penumbra, the eclipse appears as partial. When an annular eclipse takes place with the Earth at perihelion and the Moon at apogee, the negative shadow attains its greatest width, as much as 230 miles.

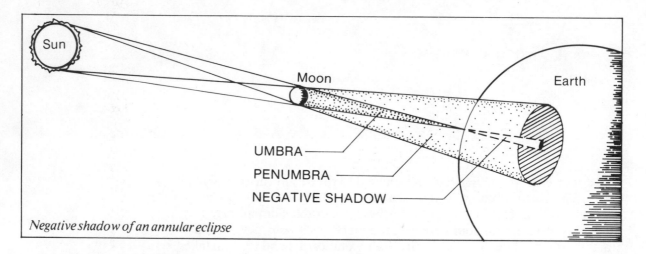

Negative shadow of an annular eclipse

Most eclipses occur when the Sun and Moon are somewhere in between their closest and farthest points. If an eclipse is not a partial one, the relative effect of the distance is calculated to determine whether the type is annular or total.

A fourth type of eclipse has both annular and total phases. Sometimes called a *central eclipse*, it starts out as annular, then becomes total, and finally reverts to annular, all in the same sweep of the shadow across the Earth. This rare type of eclipse occurs when the shadow cone of the umbra comes to a point right at the Earth. In the middle part of the eclipse, near noontime, the umbra just barely touches the Earth. The path of totality is very narrow. During the earlier and later phases of the eclipse, the umbra is not quite long enough to reach the points in the path around either side of the Earth. These *annular-total* eclipses account for only about one in every 25 solar eclipses.

Opposite: Diagram of the solar eclipse of July 18, 1860

HARPER'S WEEKLY.
A JOURNAL OF CIVILIZATION.

Vol. IV.—No. 185.] NEW YORK, SATURDAY, JULY 14, 1860. [PRICE FIVE CENTS.

Entered according to Act of Congress, in the Year 1860, by Harper & Brothers, in the Clerk's Office of the District Court for the Southern District of New York.

ECLIPSE OF THE SUN ON THE EIGHTEENTH JULY.

On 18th July inst. an eclipse of the sun will take place, which will be more or less visible throughout the United States and Canada. We publish below a diagram of the eclipse. The reader must bear in mind that it represents the degree of observation at New York; hence at all places north of this parallel the eclipse will be greater, while at all places south of New York it will be less than is represented in the diagram.

It is hardly necessary to observe that an eclipse of the sun is caused by the passage of the moon between the earth and the sun. The motions of the heavenly bodies being governed by fixed mathematical laws, each eclipse can be predicted with certainty. The first appearance of the eclipse of 18th inst. since the creation of the world (according to sacred chronology) was in the year A.D. 958, December 8, old style, at 10 o'clock 50 minutes forenoon, when the moon's penumbra just came in contact with the earth at the south pole; it has appeared every nineteenth year since, and at

until the expiration of 12,492 years, when it will come on again at the south pole, and go through a similar course. The velocity of the moon's shadow across the earth during the eclipse will be about 1850 miles an hour, or four times the velocity of a cannon-ball.

tween the Indian Territory and New Mexico; it will then take a northeasterly and then a southeasterly course over the earth. The umbra, or total dark shadow of the moon, will first come in contact with the earth in the Pacific Ocean, one hundred miles west of the coast of Oregon, direct-

and Labrador to Cape Chidley, which will be the most favorable position on the Continent for observing the total eclipse. It will then enter the Atlantic Ocean, passing due east until nearly south of Cape Farewell, the southern cape of Greenland, where the sun will be totally eclipsed at noon of that place; it will then take a curved line toward the southeast, passing over the north of Spain, the Mediterranean Sea, Algiers, Tripoli, Fezzan, the southwestern corner of Egypt, into Nubia, where it will leave the earth near the Red Sea, a little before the setting of the sun at that place. The path of the umbra, in which the sun will be totally eclipsed, will be only about seventy miles in width; whereas the penumbra, in which the sun will appear more or less eclipsed, will extend from the Gulf of Mexico to 20 degrees upon the opposite side of the north pole, a distance of over six thousand miles. The umbra, in its passage over the earth, makes a curved line; this is caused by the spherical form of the earth. If the earth were a flat surface, the path of the umbra would then be a straight line from northwest to southeast, making an angle with the equator of 17 degrees. At all places south of the line of total eclipse the sun's northern limb will be eclipsed; but in Europe, England, Ireland, Greenland, Iceland, and the northern part of British America, the southern limb will be eclipsed.

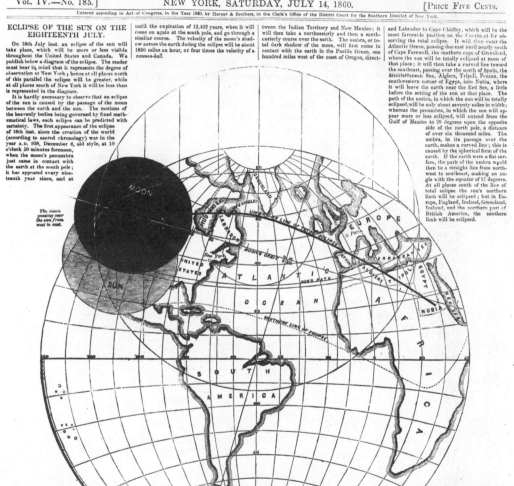

DIAGRAM OF THE ECLIPSE OF THE SUN ON JULY 18, 1860.

each return the moon's shadow passed across the earth from west to east a little farther to the north at each return, until the year 1644, March 8, old style, when the centre of the moon's shadow passed a little to the north of the earth's centre (the moon being 14 minutes 46 seconds from her descending node, which was its 38th periodical return). It has continued to appear every nineteenth year since 1644, until this eclipse, which is its sixty-first periodical return. Its next appearance will be in 1879, July 29; at 3 o'clock 23 minutes in the morning, invisible in the United States. It will also appear again in 1896, August 9. It will continue to appear every nineteenth year until the year 2274, April 25, when the moon's shadow will just touch the earth at the north pole, which will be its seventy-sixth periodical and last appearance,

THE PATH OF THE ECLIPSE.

The penumbra, or partial shadow of the moon, will first come in contact with the earth at the rising of the sun in the northern part of Texas, be-

ly west of Oregon city, and a little to the southwest of the mouth of the Columbia River. It will then pass in a northeasterly direction over British America to Hudson's Bay, near Fort York, at the mouth of Nelson's River, crossing Hudson's Bay

HOW TO USE THE DIAGRAM.

The dark circular shade on the hemisphere represents the moon passing between the earth and sun; the shaded disk represents the sun partially eclipsed by the moon. To view this eclipse, as it will appear in the heavens, July 18, at 8 o'clock 10 minutes in the morning (New York time), face the east, take hold of the top of the diagram with your left hand, and the bottom with your right, with the back of the diagram toward the eye—incline the top of the diagram toward the north at an angle of about 45°, so that the north pole of the hemisphere will point as near as possible to the North Star; with the diagram in this position look through the back toward the sun at the time of the eclipse, and you will then have a true representation of the eclipse, and the exact position of the earth, moon, and sun at the time of the greatest obscuration, and the appearance it will present viewed through a smoked

20th Century Solar Eclipses

The table on these pages lists all of the total and annular-total eclipses of the 20th century. Each eclipse is described by its date, the maximum duration of totality, and geographical areas along the central line of the eclipse (from which totality or annularity is visible).

Date		Maximum Duration (min:sec)	Maximum Width (miles)	Path of Central Line
1900 May	28	2:10	58	Mexico, United States, Spain, N. Africa
1901 May	18	6:29	149	Indian Ocean, Sumatra, Borneo, New Guinea
1903 Sep.	21	2:12	157	Antarctica
1904 Sep.	9	6:19	146	Pacific Ocean
1905 Aug.	30	3:46	123	Canada, Spain, N. Africa, Arabia
1907 Jan.	14	2:24	119	Russia, China
1908 Jan.	3	4:14	93	Pacific Ocean
1908 Dec.	23*	0:12	6	S. America, Atlantic Ocean, Indian Ocean
1909 June	17*	0:24	32	Greenland, Russia
1910 May	9	4:14	†	Antarctica
1911 Apr.	28	4:58	120	Pacific Ocean
1912 Apr.	17*	0:02	1	Atlantic Ocean, Europe, Russia
1912 Oct.	10	1:55	54	Brazil, S. Atlantic Ocean
1914 Aug.	21	2:15	113	Greenland, Europe, Middle East
1916 Feb.	3	2:36	69	Pacific Ocean, S. America, Atlantic Ocean
1918 June	8	2:23	70	Pacific Ocean, United States
1919 May	29	6:50	153	S. America, Atlantic Ocean, Africa
1921 Oct.	1	1:52	189	Antarctica
1922 Sep.	21	5:58	142	Indian Ocean, Australia
1923 Sep.	10	3:37	106	Pacific Ocean, Central America
1925 Jan.	24	2:32	130	N.E. United States, Atlantic Ocean
1926 Jan.	14	4:11	92	Africa, Indian Ocean, Borneo
1927 June	29	0:50	48	England, Scandinavia, Arctic Ocean, Russia
1928 May	19			(Umbra barely touched Antarctica)
1929 May	9	5:07	122	Indian Ocean, Malaya, Philippines
1930 Apr.	28*	0:01	1	Pacific Ocean, United States, Canada
1930 Oct.	21	1:55	54	S. Pacific Ocean
1932 Aug.	31	1:45	104	Arctic Ocean, E. Canada
1934 Feb.	14	2:53	79	Borneo, Pacific Ocean
1936 June	19	2:31	83	Greece, Turkey, Russia, Pacific Ocean
1937 June	8	7:04	156	Pacific Ocean, Peru
1938 May	29	4:04	††	S. Atlantic Ocean
1939 Oct.	12	1:32	276	Antarctica
1940 Oct.	1	5:35	137	S. America, Atlantic Ocean, Africa
1941 Sep.	21	3:22	91	Russia, China, Pacific Ocean
1943 Feb.	4	2:39	146	Japan, Pacific Ocean, Alaska
1944 Jan.	25	4:09	91	S. America, Atlantic Ocean, Africa
1945 July	9	1:16	57	Canada, Greenland, Scandinavia, Russia
1947 May	20	5:14	124	Argentina, Brazil, Central Africa

* Annular-total eclipse
† Last total eclipse in *saros* series
†† First total eclipse in *saros* series

Of the four types of solar eclipses, partial eclipses are most common. The following table gives the relative frequency of each type.

Type of Solar Eclipse	Percent of All Solar Eclipses
Partial	35
Annular	33
Total	28
Annular-total	4
	100

Date		Maximum Duration (min:sec)	Maximum Width (miles)	Path of Central Line
1948 Nov.	1	1:56	53	Africa, Indian Ocean
1950 Sep.	12	1:13	90	Arctic Ocean, Russia, Pacific Ocean
1952 Feb.	25	3:09	89	Africa, Arabia, Iran, Russia
1954 June	30	2:35	96	United States, Canada, Scandinavia, Russia
1955 June	20	7:08	159	Indian Ocean, Thailand, Pacific Ocean
1956 June	8	4:44	269	S. Pacific Ocean
1957 Oct.	23			(Umbra barely touched Antarctica)
1958 Oct.	12	5:11	131	Pacific Ocean, Argentina
1959 Oct.	2	3:01	76	Atlantic Ocean, Africa
1961 Feb.	15	2:44	164	Europe, Russia
1962 Feb.	5	4:08	92	Borneo, New Guinea, Pacific Ocean
1963 July	20	1:40	63	Pacific Ocean, Alaska, Canada
1965 May	30	5:16	124	New Zealand, Pacific Ocean
1966 Nov.	12	1:57	53	S. America, Atlantic Ocean
1967 Nov.	2			(Umbra barely touched Antarctica)
1968 Sep.	22	0:40	68	Russia
1970 Mar.	7	3:28	99	Pacific Ocean, Mexico, E. United States
1972 July	10	2:36	111	Russia, N. Canada
1973 June	30	7:04	160	Atlantic Ocean, Central Africa, Indian Ocean
1974 June	20	5:08	216	Indian Ocean, Australia
1976 Oct.	23	4:46	125	Africa, Indian Ocean, Australia
1977 Oct.	12	2:37	63	Pacific Ocean, Colombia, Venezuela
1979 Feb.	26	2:52	195	N.W. United States, Canada, Greenland
1980 Feb.	16	4:08	93	Africa, Indian Ocean, India, China
1981 July	31	2:03	68	Russia, Pacific Ocean
1983 June	11	5:11	125	Indian Ocean, New Guinea
1984 Nov.	22	1:59	53	New Guinea, S. Pacific Ocean
1985 Nov.	12	1:59	††	Antarctica
1986 Oct.	3*	0:01	1	N. Atlantic Ocean
1987 Mar.	29*	0:08	3	S. Atlantic Ocean, Central Africa
1988 Mar.	18	3:46	109	Sumatra, Borneo, Philippines
1990 July	22	2:33	130	Russia, Pacific Ocean
1991 July	11	6:54	161	Hawaii, Mexico, S. America
1992 June	30	5:20	186	S. Atlantic Ocean
1994 Nov.	3	4:23	119	Bolivia, Brazil, S. Atlantic Ocean
1995 Oct.	24	2:10	49	India, S.E. Asia, Indonesia
1997 Mar.	9	2:50	231	Russia, Arctic Ocean
1998 Feb.	26	4:08	95	Pacific Ocean, Venezuela, Atlantic Ocean
1999 Aug.	11	2:23	70	Central Europe, Middle East, India

* Annular-total eclipse
†† First total eclipse in *saros* series

Repetition of Eclipses

Where a total eclipse is visible depends mainly on three factors. First, the longitude of the eclipse path is determined by which part of the Earth happens to be facing toward the Sun and the Moon when the eclipse occurs. Second, the latitude of the eclipse path is a function of how close the Moon is to a node when the eclipse occurs. The closer to a node, the nearer to the equator will the umbra pass across the Earth. And third, the area covered by totality depends on the width (and length) of the path. The wider the path, the more locations experience totality. Astronomers use these and other factors to compute the precise path and timing for a given eclipse.

The repetition of eclipses in time follows a definite pattern. The *saros* is the best example of an eclipse cycle. Recall that after 18 years and 11⅓ days, a new Moon and a lunar node return almost exactly to their former alignment, and the eclipse is repeated. What's more, both eclipses are likely to be of the same time. This is a result of the close coincidence between the *saros* and 239 anomalistic months (239 x 27.5545 = 6,585.54 days). When the *saros* repeats, the Moon will be almost the same distance from the Earth as before. If the Moon is farther away, the eclipses will be annular; if the Moon is closer, the eclipses will be total. Also, because the eclipses take place at the same time of the year (only 11 day's difference), the Earth-Sun distance is almost the same. This almost exact resonance between the eclipse year, the synodic month, and the anomalistic month, coupled with the *saros* being close to an even number of years, results in remarkable cycles centuries long.

The progression of eclipses through a complete *saros* takes some 1,200 years. Recall that the path of each successive eclipse in a series sweeps a little farther in the same direction toward one of the poles of the Earth. The direction depends on which node is in alignment at the eclipse in a series. All the eclipses in a given series occur at either the ascending or descending node. If the eclipses take place at the ascending node, the path will be a little farther south every 18 years and 11⅓ days; if at the descending node, the eclipse tracks progress north. The *saros* series that includes the February 26, 1979, eclipse began ten centuries ago at the descending node. On May 27, 993, the Moon's penumbra barely grazed Antarctica. This partial eclipse was repeated on June 7, 1011, but this time more of the Earth was covered by the partial shadow. Each time the *saros* repeated, the Moon's shadow was a little farther north. On August 11, 1059, the first annular eclipse of the series occurred. These continued every 18 years until June 8, 1564, the date of the first total eclipse in this *saros* series. The total eclipses continue until the last one on March 30, 2033. After that, partial eclipses finish the series. The 71st and final eclipse of the *saros* cycle will be visible only as a partial eclipse from Arctic regions on July 7, 2195.

A Complete *Saros* Series

Date	Type	Date	Type
933 May 27	Partial	1582 June 20	Total
951 June 7	Partial	*1600 July 10	Total
969 June 17	Partial	1618 July 21	Total
987 June 28	Partial	1636 Aug. 1	Total
1005 July 9	Partial	1654 Aug. 12	Total
1023 July 20	Partial	1672 Aug. 22	Total
1041 July 30	Partial	1690 Sep. 3	Total
1059 Aug. 11	Annular	1708 Sep. 14	Total
1077 Aug. 21	Annular	1726 Sep. 25	Total
1095 Sep. 1	Annular	1744 Oct. 6	Total
1113 Sep. 11	Annular	1762 Oct. 17	Total
1131 Sep. 23	Annular	1780 Oct. 27	Total
1149 Oct. 3	Annular	1798 Nov. 8	Total
1167 Oct. 14	Annular	1816 Nov. 19	Total
1185 Oct. 25	Annular	1834 Nov. 30	Total
1203 Nov. 5	Annular	1852 Dec. 11	Total
1221 Nov. 15	Annular	1870 Dec. 22	Total
1239 Nov. 27	Annular	1889 Jan. 1	Total
1257 Dec. 7	Annular	1907 Jan. 14	Total
1275 Dec. 18	Annular	1925 Jan. 24	Total
1293 Dec. 29	Annular	1943 Feb. 2	Total
1312 Jan. 9	Annular	1961 Feb. 15	Total
1330 Jan. 19	Annular	1979 Feb. 26	Total
1348 Jan. 31	Annular	1997 Mar. 9	Total
1366 Feb. 10	Annular	2015 Mar. 20	Total
1384 Feb. 21	Annular	2033 Mar. 30	Total
1402 Mar. 4	Annular	2051 Apr. 11	Partial
1420 Mar. 14	Annular	2069 Apr. 21	Partial
1438 Mar. 25	Annular	2087 May 2	Partial
1456 Apr. 4	Annular	2105 May 14	Partial
1474 Apr. 16	Annular	2123 May 25	Partial
1492 Apr. 26	Annular	2141 June 4	Partial
1510 May 8	Annular-total	2159 June 16	Partial
1528 May 18	Annular-total	2177 June 26	Partial
1546 May 29	Annular-total	2195 July 7	Partial
1564 June 8	Total		

*Begin dates from Gregorian calendar

The repetition of eclipses follows very regular patterns in time. Eclipse seasons and *saros* cycles come and go like clockwork. The repetition of eclipses at a given place on the Earth, however, does not seem to follow any discernible cycle. The accompanying map shows all total eclipse paths across the U.S. and Canada since 1878. For example, Goldendale, Washington, is in the path of totality for the eclipses in 1918 and 1979. Yet many parts of North America have not had a total eclipse in the last hundred years.

Partial phases of solar eclipses can be seen about every 2½ years from the same spot. The best estimate for total eclipses is to say they recur at the same location about every 360 years on the average. This figure is based on the average width of eclipse paths, the total surface area of the Earth, and the overall frequency of total eclipses. But the actual facts vary, sometimes widely, from this estimate. The table below helps illustrate the apparent random nature of the recurrence of eclipses at the same place. The examples were chosen, not to prove any lack of pattern, but to present the flavor of the variation involved.

Location	Dates of Consecutive Total Eclipses	Years in Interval
London	Oct. 29, 878 A.D. — Apr. 22, 1715 A.D.	837
Jerusalem	Sep. 30, 1131 B.C. — July 4, 336 B.C.	795
Great Pyramid of Egypt	Apr. 1, 2471 B.C. — June 29, 2159 B.C.	312
Stonehenge	May 8, 1169 B.C. — May 7, 1066 B.C.	103
Yellowstone National Park	July 29, 1878 A.D. — Jan. 1, 1889 A.D.	11
Tomb of Tutankhamun	May 31, 957 B.C. — May 22, 948 B.C.	9
Lake Okechobee, Florida	Aug. 19, 2259 A.D. — Dec. 22, 2261 A.D.	2½
Southern New Guinea	June 11, 1983 A.D. — Nov. 22, 1984 A.D.	1½

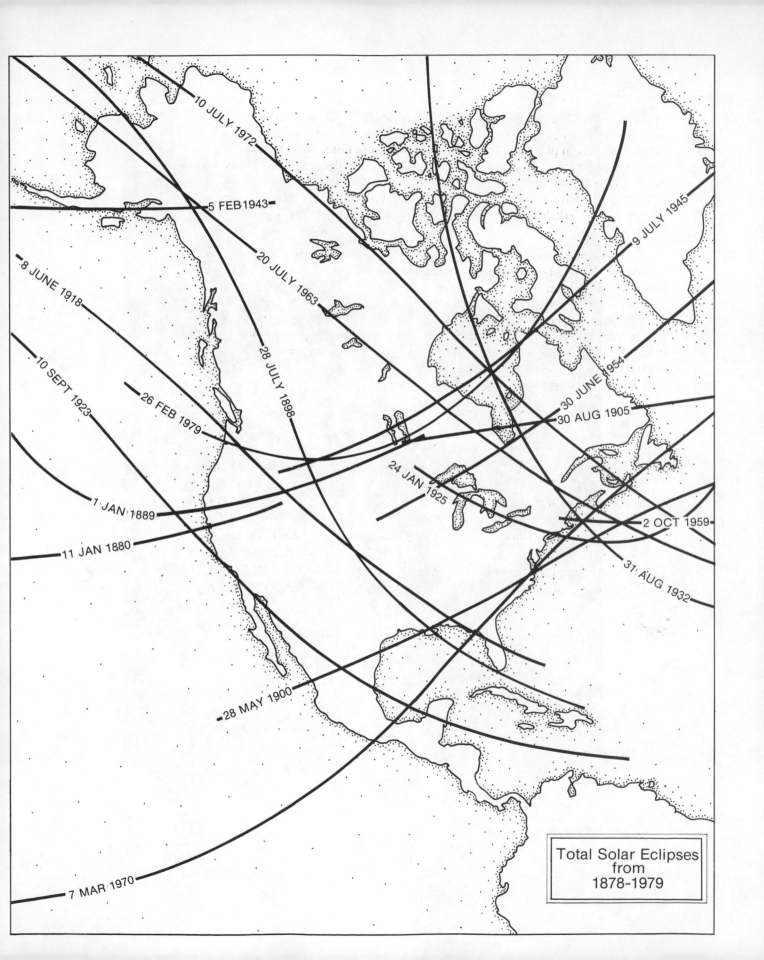

10 JULY 1972

5 FEB 1943

20 JULY 1963

8 JUNE 1918

28 JULY 1898

9 JULY 1945

1954

30 JUNE

30 AUG 1905

10 SEPT 1923

26 FEB 1979

24 JAN 1925

1 JAN 1889

2 OCT 1959

11 JAN 1880

31 AUG 1932

28 MAY 1900

7 MAR 1970

Total Solar Eclipses
from
1878-1979

Chapter 3

How To Observe an Eclipse

The spectacular sight of a total solar eclipse is for most of us a once-in-a-lifetime event. Unless you're an astronomer or an avid eclipse follower, you'll probably get only one chance to see it. It's estimated that only one in a thousand people ever experiences totality. This wondrous spectacle of the *complete* halo around the Sun can't be seen under any other earthly circumstances. (Astronomers are able to observe part of the corona without an eclipse using a coronagraph, a kind of telescope invented in 1931.) The darkness of the Moon's shadow adds to the drama of the few short minutes the corona is visible. It's simply a matter of being in the right place at the right time — and knowing what to look for.

The time and location of each eclipse, of course, is different. (These details are published a year or two in advance by the U.S. Naval Observatory.) But the observation site considerations and the viewing techniques are essentially the same for all total solar eclipses. The eclipse of February 26, 1979, is used as the example in this chapter to illustrate how to observe an eclipse.

Opposite: Path of the Moon's shadow on the Earth during the "New Year's Day Eclipse" of January 1, 1889. The total eclipse on February 26, 1979, comes exactly five *saros periods later.*

The Path of Totality

Early in the morning of February 26, 1979, as seen from a point a thousand miles off the Pacific coast of the state of Washington, the Sun rises. But instead of the familiar orange ball coming up over the horizon, the blackened disk of the eclipsed Sun appears. The umbra, the complete shadow of the Moon, makes its first grazing contact with the edge of the Earth at sunrise and the total eclipse begins. The shadow races eastward at thousands of miles per hour toward North America, where the Sun is already up on this Monday morning.

At 8:12 a.m. the umbra first touches land on the Oregon coast. The shadow sweeps along a 170-mile-wide path blotting out the Sun for up to 2½ minutes in southern Washington and northern Oregon. As the path continues across Idaho and Montana, the shadow little by little slows its movement across the land and also gets slightly wider. (Here the eclipse is occurring closer to local noon.) It reaches its maximum width, 195 miles, near the U.S.-Canadian border separating Saskatchewan and North Dakota. The path of the shadow gradually turns northeast into Manitoba. It reaches maximum duration, 2 minutes and 52 seconds, at a point just east of Lake Winnipeg.

By now the eclipse is about half over. At its slowest point, the shadow is moving about 1,700 miles per hour. It begins picking up speed as it moves across Ontario and Hudson Bay into the afternoon. The path narrows slightly as the umbra races across Northeastern Canada toward its conclusion. Finally, at sunset in Northern Greenland, the total eclipse ends. The Moon's shadow leaves the Earth and continues its course through space.

The passage of the Moon's umbra over the Earth, from sunrise in the Pacific to sunset in Greenland, takes about an hour and a half for this eclipse. During that time it travels almost 5,000 miles across the surface of the Earth. That's an average speed of more than 3,000 miles per hour. The area covered by the path of totality amounts to approximately 850,000 square miles. That's less than one half of one percent of the total surface area of the planet. Millions of people live in the path, which includes two major cities: Portland, Oregon, and Winnepeg, Manitoba. The partial phases of the eclipse are visible over virtually the entire North American continent. (The Moon's penumbra misses much of Alaska, the extreme northern parts of Canada and Greenland, and all of Panama.)

The map on the opposite page illustrates the extent of the eclipse for North America. The position of the umbra is shown at ten-minute intervals along the path of totality. (The umbra's shape is elliptical because the shadow cone strikes the Earth at an oblique angle.) The solid lines running east-west show what percent of the Sun is obscured at maximum eclipse throughout North America. The farther away from the path of totality the less of the Sun's disk is blocked out. The dotted lines show the time of maximum eclipse for both the total and partial phases.

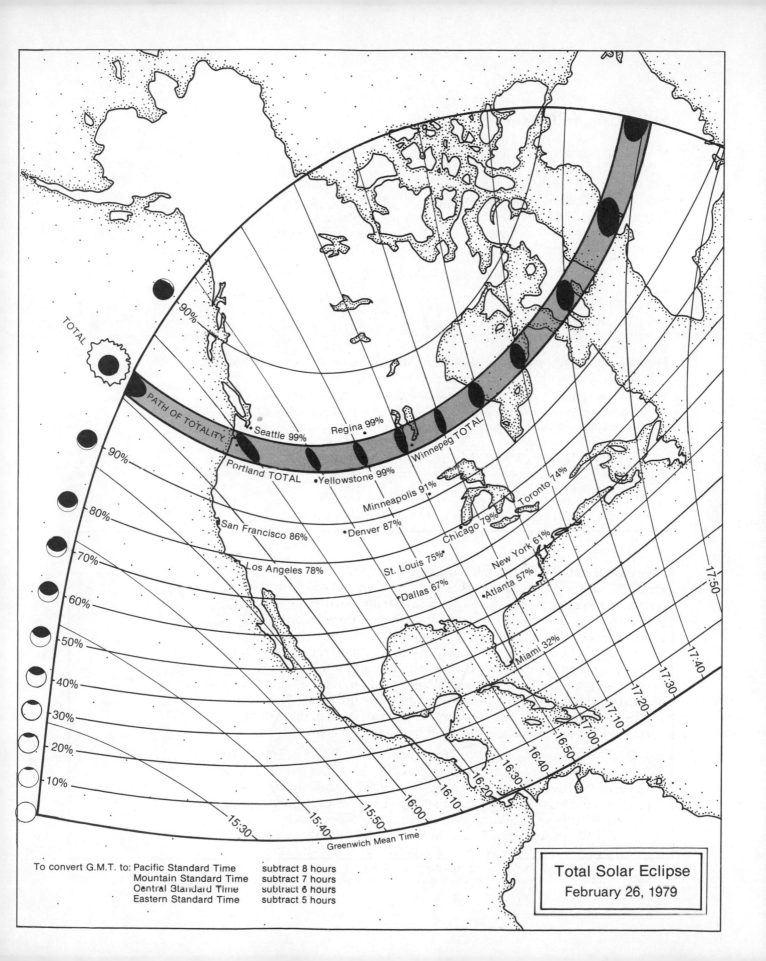

TOTAL

PATH OF TOTALITY

90%

90%

80%

70%

60%

50%

40%

30%

20%

10%

• Seattle 99%   Regina 99%

Portland TOTAL   Winnepeg TOTAL
            • Yellowstone 99%

San Francisco 86%   Minneapolis 91%
                    • Denver 87%   Toronto 74%
                                Chicago 79%•
Los Angeles 78%   St. Louis 75%•   New York 61%
                                        Atlanta 57%•
                    •Dallas 67%

Miami 32%

15:30   15:40   15:50   16:00   16:10   16:20   16:30   16:40   16:50   17:00   17:10   17:20   17:30   17:40   17:50

Greenwich Mean Time

To convert G.M.T. to: Pacific Standard Time   subtract 8 hours
                    Mountain Standard Time   subtract 7 hours
                    Central Standard Time   subtract 6 hours
                    Eastern Standard Time   subtract 5 hours

Total Solar Eclipse
February 26, 1979

Selecting an Observation Site

The partial phases of a total solar eclipse are visible over a wide area (most of North America on February 26, 1979). But only within the path of totality can you see the spectacular and striking effects. The difference between experiencing a total eclipse and a partial eclipse is, literally, "the difference between night and day." Those who live within the path or take the opportunity to travel there have the chance to be rewarded with one of the most fleeting and beautiful visions of Nature's grandeur.

Your choice of a site within the path should be guided by three main factors:
   (1) Duration of totality;
   (2) Unobstructed view of the Sun; and
   (3) Good chances for clear skies.

The map on the next two pages shows the path of totality on February 26, 1979, from the Pacific coast to **Winnepeg**. For any given locale, a point nearer the central line of the eclipse has more time in totality; this is because the Moon's shadow, which forms an ellipse on the surface of the Earth, is wider nearer the center of the path. The accompanying diagram shows this variation across the path. If you are located just within the path, totality will not last very long — less than a minute. However, the "edge phenomena" of a total eclipse (Baily's beads, **diamond-ring** effect, and view of the chromosphere) will last longer there.

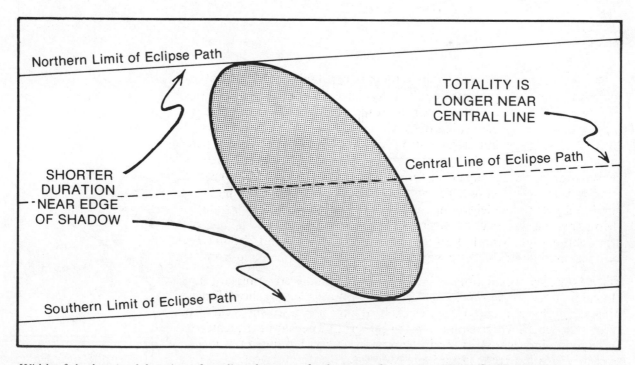

Width of shadow (and duration of totality) decreases farther away from central line of eclipse path.

To ensure an unobstructed view of the eclipse you need to know approximately where the Sun will be in the sky. You don't want any trees or mountains, for example, blocking your view. An easy way to find this out is to look for the Sun from your vantage point a day or two before the eclipse at the same time of day totality occurs. This will tell you if anything is in your way. But if this isn't practical, you can use the map on the next two pages to determine the Sun's position at total eclipse.

For this eclipse, the Sun is low in the sky and more toward the east earlier in the morning; closer to noon it's higher and more directly south. The compass direction (azimuth) to look for the Sun is shown on the map at ten-minute intervals along the path. The Sun's angle above the true horizon (altitude) is also shown. It's important to realize that the angle of altitude (for example, 12 degrees in Portland, Oregon) is measured from the true or level horizon. This doesn't account for any hills or mountains visible from your location. If you're on a hill or tall building with a good view of the west, you may also get a chance to see the approach of the Moon's shadow as it races toward you over the Earth.

Roses have thorns, and silver fountains mud;
Clouds and eclipses stain both Moon and Sun.

Shakespeare, Sonnet XXXV

The third factor in choosing a site is the weather. Unless you fly above the clouds to observe an eclipse, you'll always have to take some risk on the weather. But there is a lot you can do to optimize your chances to see the eclipse. The meteorological term for cloudy skies is "sky cover." It is measured in 10 percent increments from 0 percent to 100 percent (0 percent sky cover means clear, 100 percent is overcast). Weather Bureau records will show the average sky cover for different places along the path at the time of year of the eclipse. The weather prospects for the February 26, 1979, eclipse vary widely along the path. In the Pacific Northwest and Montana, the sky cover has averaged between 65 percent and 85 percent for February and March. Farther east in North Dakota and Canada, the chances are better: the sky cover ranges between 35 percent and 50 percent.

These sky cover predictions are only general estimates covering large areas. Local weather conditions can be very different for places a short distance apart. You'll want to avoid places likely to have fog; also, stay away from mountain ridges where clouds tend to gather. But perhaps the greatest asset in finding clear skies for an eclipse is mobility. Driving a few miles to a clearer location at the last minute could save the day for you.

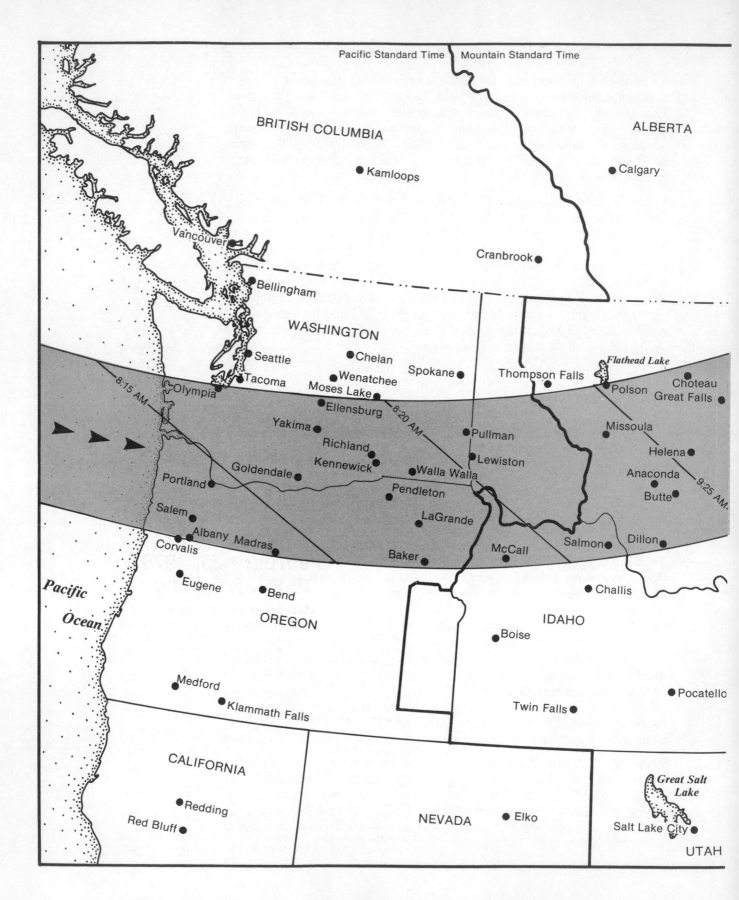

Pacific Standard Time   Mountain Standard Time

BRITISH COLUMBIA

ALBERTA

● Kamloops

● Calgary

Vancouver

Cranbrook ●

● Bellingham

WASHINGTON

8:15 AM

● Seattle   ● Chelan
                                ● Spokane   Thompson Falls ●   *Flathead Lake*
● Tacoma   ● Wenatchee   ● Polson   Choteau
Olympia   Moses Lake ●   Great Falls
                ● Ellensburg   8:20 AM
                                                                        ● Missoula
                ● Yakima   ● Pullman
                ● Richland   ● Lewiston   Helena ●
                ● Kennewick   Anaconda ●
                        ● Walla Walla   Butte ●   9:25 AM
● Goldendale   ● Pendleton
Portland ●   Salmon ●   Dillon ●
● Salem   ● LaGrande
● Albany Madras ●
Corvalis   ● Baker   McCall ●

                                                                        ● Challis

Pacific   ● Eugene   ● Bend
Ocean   OREGON   IDAHO

                                                        ● Boise

                ● Medford
                ● Klammath Falls   Twin Falls ●   ● Pocatello

        CALIFORNIA   *Great Salt
                                                                Lake*

        ● Redding   NEVADA   ● Elko
Red Bluff ●   Salt Lake City ●

                                                                        UTAH

78

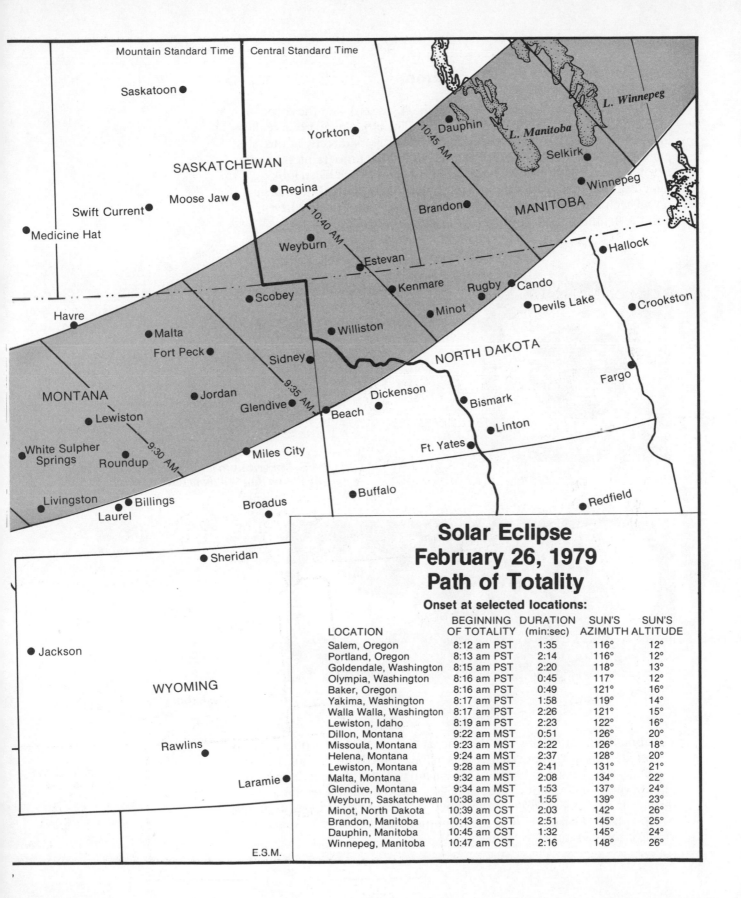

Mountain Standard Time   Central Standard Time

Saskatoon

Yorkton

SASKATCHEWAN

L. Winnepeg

L. Manitoba

Dauphin

Selkirk

Winnepeg

Moose Jaw   Regina

Brandon   MANITOBA

Swift Current

Medicine Hat

Weyburn

Estevan

Hallock

Scobey

Kenmare   Rugby   Cando

Minot   Devils Lake

Crookston

Havre

Malta

Fort Peck

Williston

NORTH DAKOTA

Sidney

Dickenson

Fargo

MONTANA

Jordan

Glendive   Beach

Bismark

Lewiston

Linton

White Sulpher Springs

Roundup

Miles City

Ft. Yates

Livingston   Billings

Laurel   Broadus

Buffalo

Redfield

Sheridan

10:45 AM

10:40 AM

9:35 AM

9:30 AM

Solar Eclipse
February 26, 1979
Path of Totality

Onset at selected locations:

LOCATION	BEGINNING OF TOTALITY	DURATION (min:sec)	SUN'S AZIMUTH	SUN'S ALTITUDE
Salem, Oregon	8:12 am PST	1:35	116°	12°
Portland, Oregon	8:13 am PST	2:14	116°	12°
Goldendale, Washington	8:15 am PST	2:20	118°	13°
Olympia, Washington	8:16 am PST	0:45	117°	12°
Baker, Oregon	8:16 am PST	0:49	121°	16°
Yakima, Washington	8:17 am PST	1:58	119°	14°
Walla Walla, Washington	8:17 am PST	2:26	121°	15°
Lewiston, Idaho	8:19 am PST	2:23	122°	16°
Dillon, Montana	9:22 am MST	0:51	126°	20°
Missoula, Montana	9:23 am MST	2:22	126°	18°
Helena, Montana	9:24 am MST	2:37	128°	20°
Lewiston, Montana	9:28 am MST	2:41	131°	21°
Malta, Montana	9:32 am MST	2:08	134°	22°
Glendive, Montana	9:34 am MST	1:53	137°	24°
Weyburn, Saskatchewan	10:38 am CST	1:55	139°	23°
Minot, North Dakota	10:39 am CST	2:03	142°	26°
Brandon, Manitoba	10:43 am CST	2:51	145°	25°
Dauphin, Manitoba	10:45 am CST	1:32	145°	24°
Winnepeg, Manitoba	10:47 am CST	2:16	148°	26°

Jackson

WYOMING

Rawlins

Laramie

E.S.M.

Observation Safety Precautions

The total phase of a solar eclipse, when the sky is dark and the corona is visible around the Sun, is a beautiful sight. The best way to observe the event during these few brief minutes is simply to look directly at this glimmering halo in the sky. The corona is a million times fainter than the bright disk of the Sun; there is no danger of eye damage when looking directly at the corona or the prominences during totality. Binoculars may reveal even finer detail, but most observers agree that the naked eye is the best "instrument" for viewing the full glory of the event.

← TOTAL PHASE: *Safe to view directly*

PARTIAL PHASES: *Dangerous to look directly at Sun,*
even if only a thin crescent is visible.
↓

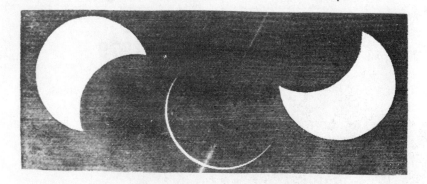

For about an hour before and after the total phase the Sun is only partially obscured. This is when it is dangerous to look directly at the Sun. Normally the Sun is too bright to look at anyway. But during these partial phases, the Sun does not appear as bright, and you may be tempted to look directly at it. DON'T DO IT! The danger of damaging your eyes does *not* depend on brightness. As long as any portion of the Sun's disk remains visible it can still cause eye damage.

The lenses of your eyes act as tiny magnifiers; if you look at the partially eclipsed Sun, its rays are focused on the retina of your eyes and can burn them. This is the same sort of thing that happens when you use a magnifying glass to focus the Sun on paper or leaves and burn a hole in them. The only difference is that it is your eyes that would be burned. Part of the danger lies in the fact that the retina is not sensitive to pain; you wouldn't even feel it happening. But a retinal burn is permanent and irreversible, producing a blank spot in the most vital part of your field of vision.

Astronomers observe the Sun directly through professionally manufactured optical filters that screen out the hazardous rays of the Sun. But unless you are trained in their use, it is not recommended that you try this method. And you're taking a big chance if you try to improvise your own filter. During the March 7, 1970, eclipse in the United States there were 145 reported cases of people who damaged their eyes by looking at the partially eclipsed Sun either directly or through sunglasses, exposed film, smoked glass, and the like. None of these homemade devices can be guaranteed safe. Play it smart and don't take any chances with your precious gift of vision.

Using smoked glass or other homemade filters, as did these observers for an eclipse in 1865, is NOT RECOMMENDED for viewing the partial phases of a solar eclipse. Serious eye damage could result.

Safe Viewing Techniques

There are some perfectly safe ways to observe the partial phases of the eclipse without looking directly at the Sun. These methods involve viewing the image of the Sun projected onto some surface; the image can be focused by having the sunlight pass through a pinhole. This is the same effect seen when the light from the partially eclipsed Sun shines through the leaves of a tree, creating tiny crescent images of the Sun. The following captioned diagrams illustrate how to build and use a simple pinhole projector. This is the safe and recommended way to observe the passage of the Moon across the face of the Sun during the *partial* phases of a solar eclipse. (If the eclipse is televised, is would also be safe to view it on the TV screen.) And don't forget: during the few minutes of totality it's OK to look directly at the Sun's corona.

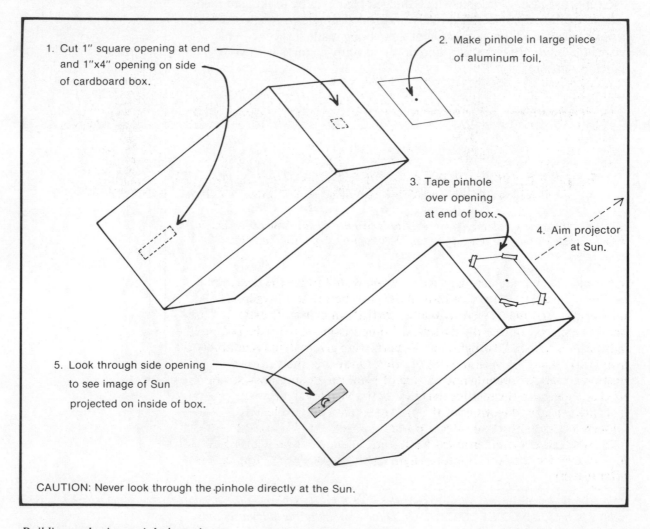

1. Cut 1" square opening at end and 1"x4" opening on side of cardboard box.

2. Make pinhole in large piece of aluminum foil.

3. Tape pinhole over opening at end of box.

4. Aim projector at Sun.

5. Look through side opening to see image of Sun projected on inside of box.

CAUTION: Never look through the pinhole directly at the Sun.

Building and using a pinhole projector

It is perfectly safe to look directly at the Sun's corona, as did these observers on July 18, 1860, during the few minutes of the total phase of a solar eclipse.

You might find it interesting to see how you would judge the degree of darkness during totality. Scientists in the past, before sensitive light-measuring instruments were available, carried out elaborate experiments to obtain some measure of the darkness. Comparisons were made to candle-light, moonlit nights, twilight, etc. Reports were given on the readability of instrument dials and various sizes of print. One experimenter even proposed that the opening and closing of plant leaves and blossoms be used as a gauge of the relative darkness in the Moon's shadow. In general, it is darker nearer the center of the path of totality and in clearer weather (clouds scatter light). You may also want to observe shadow bands. Put up a flat white sheet or screen at least three feet in diameter facing the Sun and look closely for the faint ripples of light a few minutes just before and just after totality.

Photographing a Solar Eclipse

Taking pictures of a solar eclipse can be as simple as aiming your pocket camera and pushing the button, or as complex as using sophisticated cameras mounted on telescopes driven by motors. You're not likely to get good results with the first method, but you needn't go all the way to the other extreme to produce some satisfying pictures of the corona during totality. The whole field of photography is filled with technical details. The discussion here is intended only as an overview — enough basic information to let you decide if you want to get more details from some of the sources listed as references at the end of the book.

The main problem in using a small pocket camera to photograph the corona is the short focal length of the lens; pictures taken with these types of cameras produce a very small image of the black disk of the eclipse. A camera with a lens of greater focal length will produce better results. For example, a 600mm lens will capture a circular image about ¼ inch in diameter. This is a good fit for 35mm film and gives you some leeway for errors in centering the image. It's a good idea to mount your camera on a tripod and to bracket exposure times from 1/500 second to 2 seconds; for exposures longer than 2 seconds for this size lens you should use an equatorial drive mount. (This device is explained in astro-photography references.) Choice of film seems to be a matter of personal preference; something in the range from 50 to 200 ASA should be adequate. And there is no need for any camera filters during totality.

To shoot the partial phases of the eclipse you'll need to use a filter or two to produce the equivalent of a 5.00 *neutral density* filter. Be sure you don't try to look at the Sun through these filters; they are designed for photographic use only and are not safe for your eyes. And don't look through the viewfinder at the partially eclipsed Sun. A good subject for a camera lens of shorter focus is a multiple exposure of the complete sequence of the eclipse from first to fourth contact. Use the filters for exposures of the partial phases every 5 or 6 minutes, and take one exposure of totality with the filters removed. Be sure that your camera is securely mounted and that you don't knock it out of position during the two hours or so you have it set up.

There are some good photo subjects during an eclipse other than the Sun itself. You may want to try to capture shadow bands on high-speed film using short exposures. But don't feel too disappointed if they elude your camera; they have proven very difficult to photograph, and they aren't visible at every eclipse. The crescent images of the partially eclipsed Sun seen in the shadow of a tree can make a good picture. Or you may want to try a **time-lapse** series showing the darkness of the landscape before, during, and after totality.

If you're interested in solar eclipse photography, check the references for more detailed sources of information. Or you may want to get in touch with a local amateur astronomy group; there you can swap ideas with people who have learned from experience. But whatever your attempts to photograph an eclipse, don't get so lost in your camera that you forget to look up at the corona — a sight whose beauty no film can reveal nearly as well as the human eye itself.

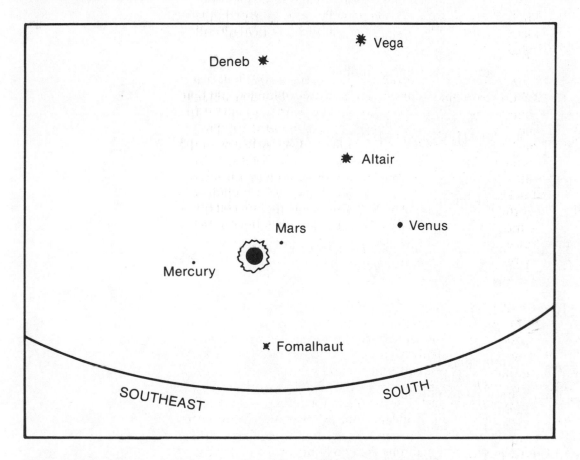

Stars and planets in the vicinity of the eclipsed Sun on February 26, 1979

Because the light of the Sun is blocked out during totality, the sky turns dark and stars and planets become visible. Although this daytime darkness lasts only a few minutes, many of the other objects in the heavens may be seen and identified. The accompanying diagram shows which stars and planets are in the sky at the eclipse on February 26, 1979. The Sun is in the constellation **Aquarius**. Mercury and Mars are in the vicinity of the eclipsed Sun on that day, but these objects are often not bright enough to be seen with the naked eye. Sometimes a bright planet, such as Venus, can be clearly seen during an eclipse. On rare occasions (as in 1882 in Egypt) a small comet may be visible.

Observation Checklist

Here is a checklist for observing an eclipse. Site selection criteria and observation equipment are summarized, followed by a list of phenomena in complete sequence for a total solar eclipse.

Site Selection:
_____ Time of eclipse
_____ Near center of path
_____ Unobscured view of Sun
_____ Sky cover prediction
_____ Provisions for last-minute mobility

Equipment:
_____ Pinhole projector
_____ Binoculars (FOR TOTAL PHASE ONLY)
_____ Cameras, film, etc.

FIRST CONTACT (Eclipse begins)
_____ Moon begins to cover Sun
_____ Crescent images of Sun
_____ Gradual darkening
_____ Shadow bands (2 or 3 minutes before totality)
_____ Baily's beads (½ minute before totality)
_____ **Approach of shadow**

SECOND CONTACT (Totality begins)
_____ Corona
_____ Prominences
_____ Chromosphere
_____ Stars and planets
_____ Darkness of landscape
_____ **Plant/animal reactions**
_____ Temperature drop

THIRD CONTACT (Totality ends)
_____ Darkness passes
_____ Baily's beads
_____ Crescent of Sun
_____ Shadow bands
_____ Gradual lightening of sky
_____ Crescent images of Sun
_____ Gradual uncovering of Sun

FOURTH CONTACT (Eclipse ends)

The Harvard Astronomical Expedition to Shelbyville, Kentucky, for the solar eclipse of August 7, 1869

Observing totality is a way of experiencing not just these brief events but a larger sense of our solar system as well. Being in the path is a unique way of becoming part of this perfect alignment of the Sun, the Moon, and the Earth. Few ever forget the experience of totality.

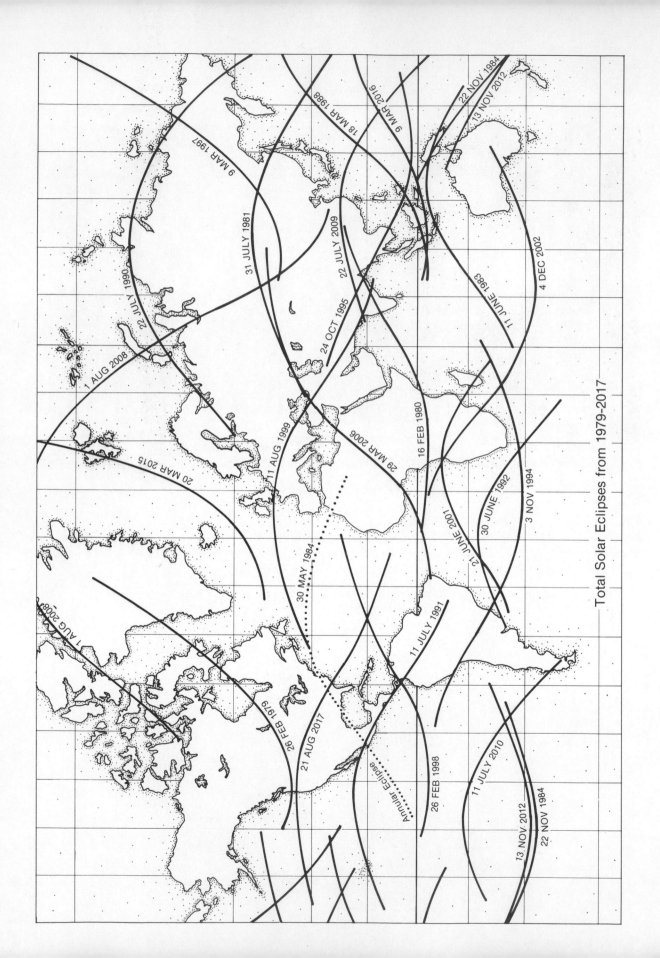

Total Solar Eclipses from 1979-2017

Epilogue

The Future

February 26, 1979, is the last time in the 20th century that the path of totality crosses any part of the continental United States or Canada. The next total solar eclipse visible from the continental United States or Canada won't occur until August 21, 2017. During this 38-year interval there are 26 total solar eclipses, but they can be seen only from other parts of the world. The accompanying map and table give the details of these eclipses. (Also shown on the map is an annular eclipse track that will cross part of the United States on the 30th of May in that fabled year, 1984.)

Thirty-eight years without a total eclipse over almost the entire North American continent is unusual. But there are also times of plenty. In the 38 years between 1594 and 1632 there were eight total eclipses visible from the U.S. or Canada. And in the 23rd century, total eclipses will be visible from the United States in the years 2245, 2252, 2254, 2259, 2261, and 2263 — a busy 18 years.

Corona seen on January 1, 1889

'Another spectacular celestial phenomenon is coming soon: Halley's Comet. Late in 1984 astronomers will detect it as a tiny speck of light on sensitive photograph plates. During the following year it will gradually brighten as it speeds toward the center of the solar system. It will make its closest approach to the Sun on February 9, 1986. For several months before and after that date it will be visible in the morning or evening skies as a bright ball of glowing gas with a long tail streaming away from the Sun.

Halley's Comet is named after English astronomer Edmond Halley. He was the first to establish a periodic pattern for a comet. He observed this comet in 1682 and noted that its orbit was similar to two earlier in 1531 and 1607. Reasoning that it was the same comet reappearing about every 76 years, Halley predicted it would return in 1759. Halley died in 1742 at age 86, too early to find out if he was right. But the comet did follow his prediction and has borne his name ever since.

Halley's Comet has appeared regularly throughout the centuries, as shown by many historical records of its sighting. Probably the most famous occurred in April 1066: historians wrote that William the Conqueror was guided by a comet in the Norman invasion of England. After its predicted return in 1759, Halley's Comet appeared again in 1835 and in 1910. This most recent appearance coincided with the death of King Edward VII in England. During that return Halley's Comet passed within 15 million miles of the Earth. This was a remarkably close call considering that the comet's orbit reaches 3.2 billion miles from the Sun, almost to the edge of the solar system.

Halley's Comet, June 7, 1910

Detail of the Bayeux Tapestry created in the late 11th century. King Harold's subjects are shown pointing to Halley's Comet (top center); to the left is a Latin inscription that translates "They marvel at the star."

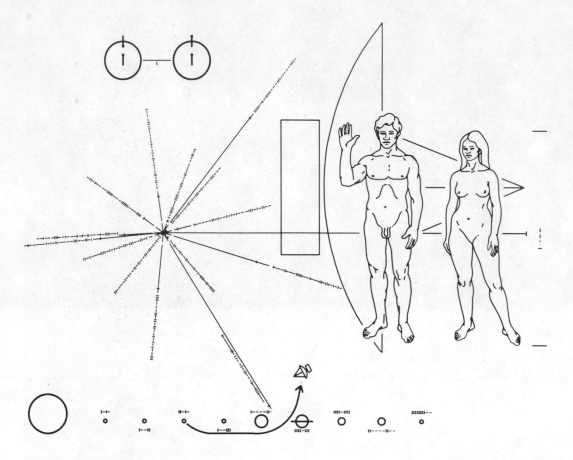

NASA's Pioneer X plaque (1972)

Shortly after Halley's Comet reaches around the Sun the Pioneer X space probe will become the first **human-made** object to leave the solar system. It will pass beyond Pluto's orbit in 1987 and enter the realm of deep space, traveling at a speed of seven miles per second. Before Pioneer X was launched in 1972 to study Jupiter, astronomers Carl Sagan and Frank Drake persuaded NASA to include in the program a unique experiment: Attached to the spacecraft is a plaque designed to identify its origin to any intelligent extraterrestial beings who might intercept it. The figures and diagrams etched on the gold plaque convey nonverbal information about the solar system, its location in the Galaxy, and the inhabitants of the third planet, Earth.

The two human figures represent typical earthlings. Their size is shown in comparison to an outline of the Pioneer X spacecraft in the background. The two circles at the top represent atoms of hydrogen, the most abundant element in the universe. The starburst pattern shows the direction and distances to specific stars. At the bottom of the plaque are the Sun and the planets. Pioneer X is shown on its path leaving Earth, swinging by Jupiter, and passing forever beyond the solar system.

William Blake's The Serpent Temple, *from* Jerusalem *(1820)*

Pioneer X and other missions like it are rapidly expanding our knowledge of the other planets in the solar system. We have come a long way from stone monuments to interplanetary space probes. Yet our fascination with the Sun, the Moon, the stars, and the planets remains. Early attempts to understand the heavens were based on astrology: the **prediction of earthly** events by the positions of the planets. It was against this background that modern astronomy has developed. As we increase our knowledge of the universe and understand more how everything is related, new patterns of reality will emerge. That is the destiny of science. Our expansion into the solar system is already bringing us exciting new views of other planets and other moons. But none of these planets or moons can produce an eclipse like we see here on Earth; this perfect matching of our lunar and solar disks is unique in the solar system. It's this remarkable coincidence in time and space that gives us this experience we call a total eclipse of the Sun.

Glossary

altitude — the angle (in degrees) above the level horizon of an object in the sky. (The object's azimuth is also needed to pinpoint its position.)

annular eclipse — a solar eclipse that occurs when the apparent size of the Moon is not great enough to completely cover the Sun. A thin ring of sunlight is seen all around the black disk of the Moon.

annular-total eclipse — a solar eclipse that has both annular and total phases. (Also called a central eclipse.)

anomalistic month — the time it takes for the Moon to travel from apogee to perigee and back again (about 27.6 days).

aphelion — the point in the Earth's orbit that is farthest from the Sun. Currently the Earth reaches aphelion in early July.

apogee — the point in the Moon's orbit that is farthest from the Earth.

ascending node — the point in the orbit of the Moon where it passes from below to above the ecliptic plane (see *node*).

Aubrey holes — the 56 chalk-filled holes (named for John Aubrey) that mark the outer ring of Stonehenge. These holes may have served as "counters" to help in predicting eclipses.

azimuth — the compass direction (in degrees) of an object in the sky. (The object's altitude is also needed to pinpoint the object's position.)

Baily's beads — the effect seen just before and just after totality when only a few points of sunlight are visible at the edge of the lunar disk.

canon — in ancient times, an historical record of events. In modern astronomy, a canon is a listing of celestial events, such as eclipses, over a period of time.

central eclipse — (See *annular-total eclipse*) In some references, a central eclipse refers to an eclipse that is either total or annular.

chromosphere — the lower atmosphere of the Sun that appears during an eclipse as a thin rosy ring around the edge of the solar disk.

corona — the upper atmosphere of the Sun that appears during an eclipse as a halo around the Sun.

contact — one of the instances when the edges of the Sun and Moon cross another during an eclipse. They are designated as first contact, second contact, third contact, and fourth contact.

descending node — the point in the orbit of the Moon where it passes from above to below the ecliptic plane (see *node*).

draconic month — the time it takes for the Moon to return to a node (about 27.2 days).

eclipse — the alignment of celestial bodies so that one is obscured, either partially or totally, by the other.

eclipse season — the period of time when the Sun is near alignment with a lunar node during which eclipses may take place. For solar eclipses, this time window of 37½ days occurs about every six months.

eclipse year — the length of time for a lunar node to return to its original alignment with respect to the Sun (about 346.6 days).

ecliptic — the plane of the Earth's orbit around the Sun. As seen from the Earth, the Sun appears to move across the ecliptic during one year.

equinox — either of the two days when the periods of daylight and darkness are of equal length. The vernal equinox is usually March 21; the autumnal equinox is usually September 23.

first contact — the beginning of a solar eclipse marked by the edge of the Moon completely passing away from the disk of the Sun.

Fourth contact — the end of a solar eclipse marked by the disk of the Moon completely passing away from the disk of the Sun.

G.M.T. — Greenwich Mean Time. The time at Greenwich, England, which is used as the basis for standard time throughout the world.

heel stone — the large upright boulder at Stonehenge that is aligned with the summer solstice sunrise.

latitude — distance on the Earth (measured in degrees) north or south of the equator.

longitude — distance on the Earth (measured in degrees) east or west from a reference line, usually the line running between the poles passing through Greenwich, England.

lunar eclipse — the passage of the Moon into the shadow of the Earth. The Sun's light is cut off either partially or totally from the Moon.

node — the two points where a tilted orbit intersects a geometrical plane. The Moon's orbit intersects the ecliptic plane at the ascending node and the descending node.

partial eclipse — an eclipse during which only the partial shadow touches the Earth (for a solar eclipse) or the Moon (for a lunar eclipse).

penumbra — the part of a shadow (as of the Moon) within which the source of light (the Sun) is only partially blocked out.

perigee — the point in the orbit of the Moon that is closest to the Earth.

perihelion — the point in the orbit of the Earth that is closest to the Sun. Currently the Earth reaches perihelion in early January.

prominence — a large-scale gaseous formation above the surface of the Sun.

regression — the movement of points in an orbit in the direction opposite from the motion of the orbiting body. For example, the Moon travels from west to east, but its nodes are regressing from east to west.

saros — the eclipse cycle with a period of 223 synodic months, or 6,585.32 days (18 years and about 11 days).

second contact — the beginning of the total phase of a solar eclipse marked by the leading edge of the Moon first completely obscuring the Sun.

shadow bands — faint ripples of light sometimes seen on flat, light-colored surfaces just before and just after totality.

solar eclipse — the passage of the Moon directly between the Sun and the Earth when the Moon's shadow is cast upon the Earth. The Sun appears in the sky either partially or totally covered by the Moon.

solstice — the day when the noontime Sun is either highest in the sky (summer solstice is June 22) or lowest in the sky (winter solstice on December 22).

spectroscope — a scientific instrument that breaks down light into its component wavelengths for measurement.

sunspot — a magnetic disturbance on the Sun that appears as a dark blotch on the surface.

synodic month — the time from one full Moon to the next (about 29.5 days).

third contact — the end of the total phase of a solar eclipse marked by the trailing edge of the Moon first revealing the Sun.

total eclipse — an eclipse during which the complete shadow touches the Earth (for a solar eclipse) or the Moon (for a lunar eclipse).

totality — the period during a solar eclipse when the Sun is completely blocked by the Moon. (Totality for a lunar eclipse is the period when the Moon is in the complete shadow of the Earth.)

umbra — a complete shadow (as of the Moon) within which the source of light (the Sun) is totally hidden from view.

zodiac — the division of the ecliptic into twelve equal parts; each of these parts or "signs" is identified by a name and symbol (for example, Sagittarius ⟋).

References

GENERAL READING

Chambers, George F., *The Story of Eclipses,* D. Appleton & Co., New York (1912). This handy little volume has much detail on historical references to solar and lunar eclipses.

Dyson, Frank, and Woolley, R. v.d. R., *Eclipses of the Sun and Moon,* Oxford Univ. Press, London (1937). This textbook on eclipses has an emphasis on the mathematics and physics of eclipses.

Flammarion, Camille, *The Flammarion Book of Astronomy,* Simon & Schuster, New York (1964). This completely updated version of a popular 19th century astronomy book has several good chapters on eclipses.

Mitchell, Samuel A., *Eclipses of the Sun*, Columbia Univ. Press, New York (fifth edition, 1951). This basic text on solar eclipses blends the author's personal experiences with the history and science of eclipses.

Todd, Mabel L., *Total Eclipses of the Sun*, Little, Brown, & Co., Boston (1900). This popular treatment of the subject was produced in anticipation of the May 28, 1900, eclipse in the United States.

National Geographic has included a number of articles on solar eclipses. See the issues of August 1970, November 1963, March 1949, September 1947, September 1937, February 1937, and November 1932.

Sky and Telescope is another useful source of information. Hardly an issue passes without some treatment of eclipses and related subjects.

ECLIPSE DATA

American Ephemeris and Nautical Almanac, U.S. Government Printing Office, Washington, D.C. 20402. This annual volume contains (among much other information) the details of all eclipses for the year.

Kudlek, Manfred, and Mickler, Erich H., *Solar and Lunar Eclipses of the Ancient Near East,* Verlag Butzon & Bercker Kevelaer, Hamburg (1971). This book gives eclipse data and maps from 3000 B.C. to the year 0 for important historical places in the ancient Near East.

Meeus, J., Grosjean, C.C., and Vanderleen, W., *Canon of Solar Eclipses,* Pergamon Press, Oxford (1966). This canon contains the worldwide data and maps of all solar eclipses between 1898 and 2510.

Oppolzer, Theodor R. von, *Canon of Eclipses*, Dover Publications, New York (1962). This is a reprint of Oppolzer's classic of 1887; it presents data and maps for eclipses from 1207 B.C. to 2161 A.D.

SOLAR ECLIPSE PHOTOGRAPHY

Paul, Henry, *Outer Space Photography,* American Photographic Book
Publishing Company (1960). Includes a chapter on eclipse photography.

Astrophotography Basics, Kodak Customer Service Pamphlet Number
AC-48 (1978). This concise and informative guide has much good
information on eclipse photography.

STONEHENGE

Hawkins, Gerald S., *Stonehenge Decoded*, Doubleday, Garden City
(1965). This is the original account of the discovery that Stonehenge may
have been used to predict eclipses.

Hoyle, Fred, *On Stonehenge,* W.H. Freeman & Co., San Francisco (1977).
This book delves deeper into the astronomical aspects of this ancient
monument.

Index

Acknowledgments

Ann Ronan Picture Library and Royal Astronomical Society, *23*

Bettman Archive, *6*

Crown Copyright; reproduced with permission of the Controller of Her Majesty's Stationery Office, *8, 10, 54*

Dennis diCicco, *Sky and Telescope, 37*

High Altitude Observatory, National Center for Atmospheric Research, *3, 39, 40*

Houghton Library, Harvard University, *24*

Lick Observatory, *2, 4, 38, 90-91*

NASA, *92*, front & back cover

Seattle Art Museum, *16*

Seattle Times, *42*

Special Collections Division, University of Washington Libraries, *45, 47, 57, 60, 61, 91, 93*

About the Author

Bryan Brewer, born December 6, 1946, has resided in the Pacific Northwest since 1972. During that time he has used his skills as a writer and instructor in such diverse areas as computer science, the energy of form, and the psychology of consciousness. He and his wife and son are building a log house near Mt. Rainier, Washington, and hope to make their home there eventually.